The Ranque Hilsch Vortex Tube Demystified

André Kaufmann

The Ranque Hilsch Vortex Tube Demystified

Understanding the Working Principles of the Vortex Tube

Springer

André Kaufmann
Ravensburg Weingarten University of
Applied Sciences
Weingarten, Germany

ISBN 978-3-030-89765-9 ISBN 978-3-030-89766-6 (eBook)
https://doi.org/10.1007/978-3-030-89766-6

This Springer imprint is published by the registered company Springer Nature Switzerland AG
The registered company address is: Gewerbestrasse 11, 6330 Cham, Switzerland

Publisher

Preface

This booklet resulted from a question which one of my colleagues asked me: "Is it possible to cool a server (computer) at reasonable cost with a Ranque-Hilsch Vortex Tube?" Honestly, at that time I did not know the answer. However, the question inspired me to find out about its possible working mechanisms and to write it down in the form of this booklet. Today, I would tell my colleague that he would be better off with an air conditioner.

Rather than providing rigorous proofs, the emphasis in this book is on providing illustrative examples, many sketches and illustrations, and some handwaving arguments to qualitatively describe vortex tube behaviour and its effects.

So, this booklet is for my colleague who has chosen to remain anonymous and all those who are intrigued by the working principle of the vortex tube.

In this booklet the reader will learn about:

- Basic vortex tube geometries
- Thermodynamics of the vortex tube
- Flow structure inside the vortex tube
- Why viscosity plays such an important role in the balance of energy

This book has benefited greatly from the corrections by A.P. Weiss and J. Sorebo. Thank you.
Special thanks go to all those who gave me the time to work on this topic and to write this book. In addition to the family, this also includes support from the university in the form of a research semester.
If you find a typo please let me know: Andre.Kaufmann(a)rwu.de.
Enjoy reading,

Baienfurt, Germany André Kaufmann

Contents

Chapter 1
Introduction to the Ranque–Hilsch Vortex Tube

> The first principle is that you must not fool yourself and you are the easiest person to fool.
>
> Richard P. Feynman

Since the process of temperature separation by the vortex tube was first patented and published by Georges J. Ranque in 1933, the underlying physics have intrigued scientists and engineers. The device was continuously refined and found use in a variety of applications ranging from the cooling of milling tools to the cooling of clothing for firefighters. The vortex tube bifurcates a high pressure gas flow between a cold and a hot mass gas fraction. Effects of mixing are well known, but the effect of separating flow into cold and and hot is unique to the device.

In this process, the geometric simplicity and the absence of any mechanically moving parts are outstanding. Nothing more but a compressed air supply is needed to generate a hot and cold stream gas flow. The idea of this booklet is to take away the burden from the reader of searching through all of the available literature to get an idea of the working principle.

The vortex tube is often dubbed the Ranque–Hilsch Vortex Tube (RHVT) after the discovery and patent by Ranque and the refinement and first systematic experiments by Rudolf Hilsch [12]. Throughout the book we will therefore refer to the vortex tube by the abbreviation RHVT.

1.1 Working Principle of the Vortex Tube

The device is remarkably simple. Air from a high pressure reservoir enters a tube with a small diameter tangentially and imposes a strong rotating motion in the tube. The compressed gas evades partially by an annular gap at the outer diameter

A, Kaufmann, *The Ranque Hilsch Vortex Tube Demystified*,
https://doi.org/10.1007/978-3-030-89766-6_1

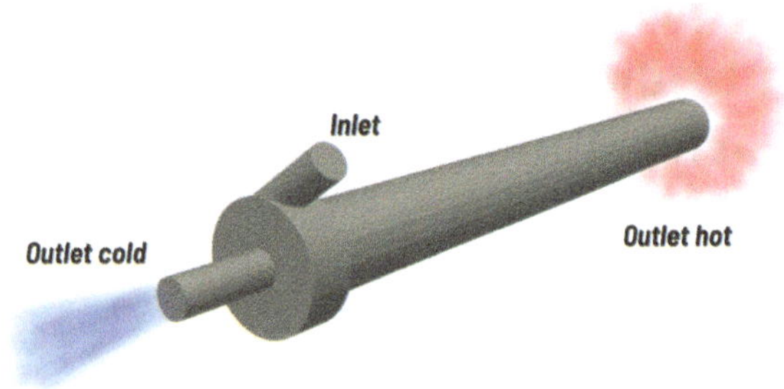

Fig. 1.1 Sketch of a counter-flow vortex tube

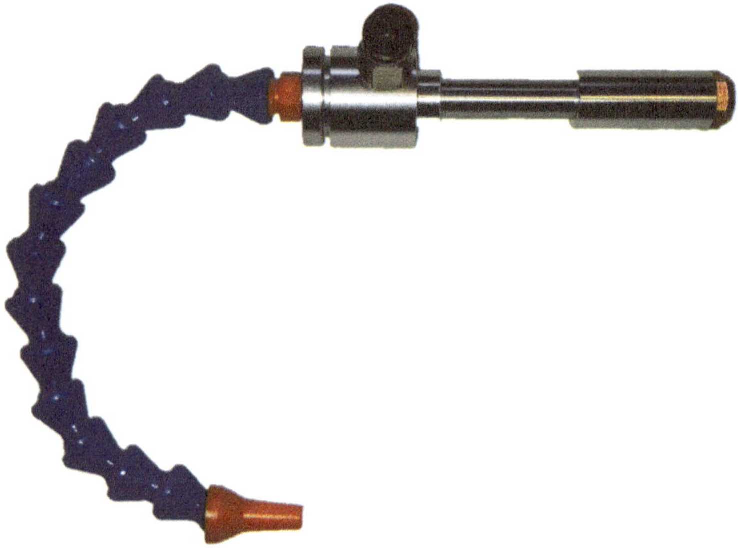

Fig. 1.2 Commercial vortex tube for spot cooling of milling tools

of the tube with higher temperatures than entering the reservoir through the inlet nozzle. The remaining fraction of the compressed gas leaves the tube with lower temperature in the centre exit. The temperature difference between the hot and the cold exit can easily span 40 K. Due to the high velocities, the air leaving the device generates significant noise if no mufflers are used. This effect supports the argument that acoustic streaming is the cause of temperature separation. A sketch of the RHVT is given in Fig. 1.1. Typically inner tube diameters are in the order of $d = 5-100\,\text{mm}$ and the length to diameter ratio is typically of the order of $l/d = 10-20$. Figure 1.2 shows a commercial vortex tube for spot cooling of milling tools.

The magnitude of the temperature difference depends on the mass fractions of the mass fluxes leaving the hot and cold exit and the pressure ratio of the reservoir to the exit pressure. The mass fraction is often adjusted by a conical plug at the hot end which serves to vary the exit cross section.

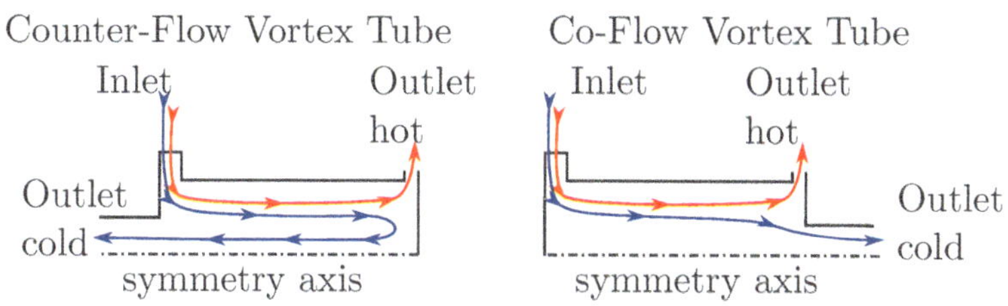

Fig. 1.3 Counter- and co-flow RHVT geometries

1.2 Vortex Tube Geometries

There are two basic geometry types: the counter-flow and the co-flow geometry as shown in Fig. 1.3. The most common is the counter-flow geometry. In this case the cold gas exits on the axis of symmetry close to the tangential inlet. The hot gas leaves the tube through the annular gap at the opposed end radially outward. The less common co-flow geometry has a centre hole in the plug on the hot side end that serves as the exit orifice for the cold mass flow. Altogether, the geometries have similar performance characteristics.

The RHVT may have more than one inlet nozzle, see, for instance, Shamaddini [19]. Some vortex tubes have grooves on the inlet to establish a homogeneous circumferential velocity.

The geometry is typically characterised by both dimensional and non-dimensional parameters. Those include tube diameter, length to diameter ratio, a number of inlet nozzles and nozzle geometries, inlet and outlet cross sections. Numerous publications describe how changes to these parametric values can be used to study their impact on energy separation. The set of literature on this topic is vast, so only the most relevant sources will be cited here.

1.3 Literature on the RHVT

The device has garnered significant attention in the scientific community and is the subject of numerous publications in respected journals. In some review papers [8, 8, 9, 20, 22, 25], a summary of the experimental work and the CFD computations is given.

Numerous attempts were made to establish non-dimensional numbers that allow a correlation for the performance of the RHVT, see Piralishvili [16], for instance. But only the expansion ratio from the pressure reservoir and the cold mass fraction have proved to be reliable correlating factors influencing the temperature separation.

1.4 Attempts to Explain the Working Mechanism

In the available literature, the vortex tube is investigated by experiment and computational fluid dynamics (CFD). These investigations have led to a number of possible explanations for the cause of the temperature separation effect. Among the most commonly cited causes in alphabetical order are:

- Acoustic streaming (Kurosaka [13])
  Sound waves can generate a steady mean flow in special geometries. Thus a transfer of energy from the sensible enthalpy to the kinetic energy is possible.
- Euler turbine effect (Polihronov and Straatman [17])
  The authors claim that the temperature decrease can be explained by the Euler turbine equation.

$$T_t - \frac{\vec{v}.\vec{\omega} \times \vec{r}}{cp} = const. \tag{1.1}$$

  Here T_t is the total temperature of gas at radial position $\vec{r}$, $\vec{v}$ is the absolute gas velocity as observed from the stationary frame of reference and $\vec{\omega}$ is the angular velocity of the system. However, there is no moving part in the system that could transfer the energy to the outside.
- Görtler vortices (Stephan et al. [21])
  In the boundary layer along a concave wall, secondary flow occurs in the form of vortices. The authors claim that Görtler vortices play an important role in the energy separation mechanism.
- Viscous effects (Van Deemter [23])
  The high swirling velocity causes strong friction in the wall boundary layers. The strong velocity gradients cause a viscous energy flux transferring kinetic energy into the boundary layer.

In the chapter on energy conservation, some strong arguments for the theory explaining the energy separation by viscous effects can be found.

1.5 Outline of the Following Chapters

In the following chapters, the reader will be guided from the macroscopic thermodynamic observations in Chap. 2 through the flow patterns in Chap. 3 and the microscopic energy balance in Chap. 4. To some extent, this follows the historical developments in research on the topic. At the same time, it allows for a natural progression, moving from the macroscopic effects to the microscopic mechanisms.

Combining the observations and theoretical results, the working principle of the RHVT is outlined in Chap. 5.

The reader is expected to be acquainted with some rudimentary knowledge of thermodynamics, especially conservation laws and some principles of fluid dynamics. Some familiarity with analytical methods is helpful to understand the derivations of the balance equations.

However the latter is not necessary to grasp the working principle concepts.

Wherever appropriate, the ideas are supported with my own illustrations and computational results.

Chapter 2
The Thermodynamic Point of View

> If you can't explain it simply, you don't understand it well enough.
>
> Albert Einstein

Before computational fluid dynamics (CFD) became a widely accessible tool, investigations on the RHVT were done by experiment. Key parameters typically measured were inlet and outlet temperatures, the expansion ratio, and the mass flow ratio of cold to total mass flow. The RHVT itself was more or less treated as a black box device.

The only possible tool for checking the plausibility of research findings was thermodynamics. In this chapter experimental results from literature are used to explain that the RHVT obeys the laws of classical thermodynamics. Some topics arising when checking the thermodynamics are examined more closely using CFD in a later chapter.

2.1 Conservation of Mass

From the thermodynamic point of view the RHVT splits a mass flow entering the tube in a mass flow with low enthalpy(cold flow) and a mass flow with high enthalpy (hot flow). The conservation of mass flow requires the sum of cold mass flow $\dot{m}_c$, hot mass flow $\dot{m}_h$ and the inlet mass flow $\dot{m}_i$ to be equal to zero.

$$\dot{m}_i + \dot{m}_c + \dot{m}_h = 0 \tag{2.1}$$

Here we follow the thermodynamic convention that the mass flow entering the control volume is always positive and the mass flow exiting the control volume is negative. Naming convention and splitting of mass flow is illustrated in Fig. 2.1.

A, Kaufmann, *The Ranque Hilsch Vortex Tube Demystified*,
https://doi.org/10.1007/978-3-030-89766-6_2

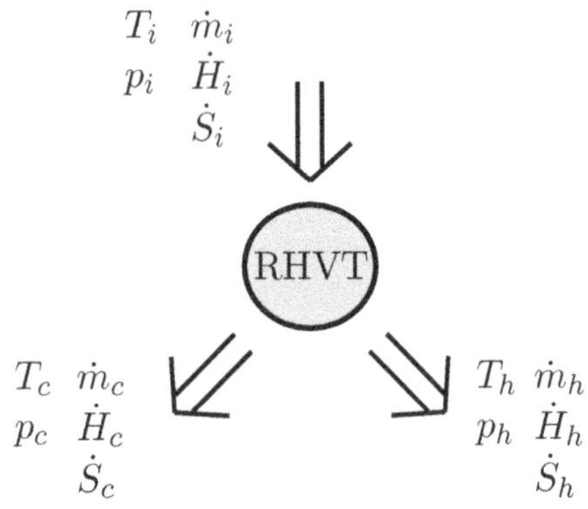

Fig. 2.1 Splitting of mass flow, enthalpy flux and entropy flux in the RHVT

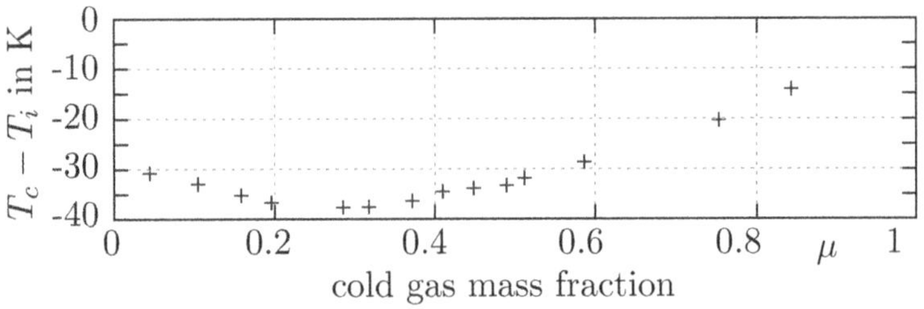

Fig. 2.2 Temperature separation as a function of mass flow for air as a working fluid [21] at inlet pressure $p_i = 5$ bar and ambient outlet pressure $p_a = 1$ bar

Experiments and numerical investigations show that the separation of the mass fluxes has a large impact on the temperature separation. Therefore it is common in the RHVT literature to characterise the flow by the cold gas mass fraction μ.

$$\mu = \frac{|\dot{m}_c|}{\dot{m}_i} \tag{2.2}$$

The cold mass fraction $\dot{m}_c$ is leaving the control volume. By thermodynamic convention it is negative. The absolute value of the mass flow is therefore used in the equation on the mass flow fraction.

The effect of mass fraction on the temperature splitting for the experiment by Stephan et al. [21] is given in Fig. 2.2. It shows that the lowest temperature at the cold end is attained at a mass flow fraction of roughly $\mu \approx 0.3$ and tends to zero for mass flow fractions of $\mu = 0$ and $\mu = 1$. This limiting behaviour can be explained with the energy balance.

2.2 Conservation of Energy

The next important factor we will explore is energy conservation. When heat transfer from the flow to the outside of the RHVT is neglected, the adiabatic energy conservation is most conveniently expressed as an enthalpy flux conservation. The enthalpy flux of the cold mass flow $\dot{H}_c$, the enthalpy flux of the hot mass flow $\dot{H}_h$ and the enthalpy flux of the inlet mass flow $\dot{H}_i$ must sum up to zero.

$$\dot{H}_i + \dot{H}_c + \dot{H}_h = 0 \tag{2.3}$$

As in the case of the mass flow rates, enthalpy flux entering the control volume is positive and enthalpy flux leaving the control volume is negative by convention. Since no mechanical work is derived from the device and no heat flux to the outside is considered, the equation above (Eq. 2.3) holistically represents how energy is conserved. The equation is valid for gaseous and liquid flow. A simplification can be made if the gas or liquid is assumed to follow a caloric perfect state equation with a fixed heat capacity c_p.

$$\dot{H} = \dot{m} c_p T \tag{2.4}$$

The equation of enthalpy conservation (Eq. 2.3) can then be expressed with the use of the cold gas mass fraction.

$$\begin{aligned} \dot{m}_i c_p T_i + \dot{m}_c c_p T_c + \dot{m}_h c_p T_h &= 0 \\ T_i &= \mu T_c + (1 - \mu) T_h \end{aligned} \tag{2.5}$$

If the cold gas mass fraction μ is known, the equation for the temperature T still has two remaining unknowns, T_c and T_h, and cannot therefore be used to explicitly calculate the temperature splitting.

However, it shows that in the limiting cases of $\mu = 0$ and $\mu = 1$ energy conservation requires that the temperature of the gas leaving the tube must equal the temperature of gas entering. The cold or hot mass fraction temperature must equal the inlet temperature T_i. As pointed out in the previous section, experiments with air as working fluid show that the maximum temperature separation is often attained at $\mu \approx 0.3$ and cold temperatures may reach $T = -20\,^{\circ}\mathrm{C} \equiv 253\,\mathrm{K}$ when starting with $T_i = 20\,^{\circ}\mathrm{C} \equiv 293\,\mathrm{K}$. Then by the equation above the hot gas temperature can be computed.

$$T_h = \frac{T_i - \mu T_c}{1 - \mu} = \frac{293\,\mathrm{K} - 0.3 \cdot 253\,\mathrm{K}}{1 - 0.3} = 310\,\mathrm{K} \equiv 37\,^{\circ}\mathrm{C} \tag{2.6}$$

Table 2.1 Comparison of temperature equation to experimental results taken from the review paper of Eiamsa-ard [9]

Year	Author	$T_h - T_i$ in K	$T_c - T_i$ in K	μ_c	$T_c - T_i$ theory in K
1951	Scheper	3.9	−11.7	0.26	−11.1
1957	Scheller	15.6	−23	0.506	−15.2
1958	Otten	40	−50	0.43	−53.0
1968	Vennos	−1	−13	0.35	1.9
1969	Bruun	6	−20	0.23	−20.1
1983	Stephan	28	−38	0.3	−65.3
1983	Amitani	15	−19	0.4	−22.5
1988	Negm	30	−42	0.38	−48.9
1996	Ahlborn	30	−27	0.4	−45.0
2005	Promvonge	25	−30	0.38	−40.8
2005	Aljuwayhel	1.2	−11	0.1	−10.8

The temperature balance equation can be checked with experimental results from the literature. This only requires the balance to be written as a temperature difference.

$$T_c - T_i = \left(1 - \frac{1}{\mu}\right)(T_h - T_i) \tag{2.7}$$

Table 2.1 shows a comparison between the experimental data of temperature difference ($T_c - T_i$) of several authors compared to the adiabatic temperature difference ($T_c - T_i$) from Eq. 2.7. This shows the correct trend for the magnitude of temperature difference. The difference between adiabatic temperature difference and experiment can be explained by the diabatic behaviour in the experiments. The dependence of temperature separation $T_c - T_i$ on the cold gas mass fraction μ is the subject of many publications. For instance, Stephan et al. [21] give an empirical correlation that fits their experimental data sets and the data of Hilsch [12].

$$\frac{T_c - T_i}{(T_c - T_i)_{max}} = 0.792 + 1.540\mu - 3.101\mu^2 + 0.815\mu^3 \tag{2.8}$$

Similar empirical correlations can be found in numerous other publications. However, this correlation does not fulfill the relation obtained from energy conservation. The enthalpy balance requires the temperature difference to be 0 in the case of $\mu = 0$ and $\mu = 1$.

The conditions are fulfilled by an empirical correlation of the following type.

$$T_c - T_i = \Delta T \left(1 + \frac{m}{n}\right)^{n/m} \left(1 + \frac{n}{m}\right) \mu^n (1 - \mu^m) \tag{2.9}$$

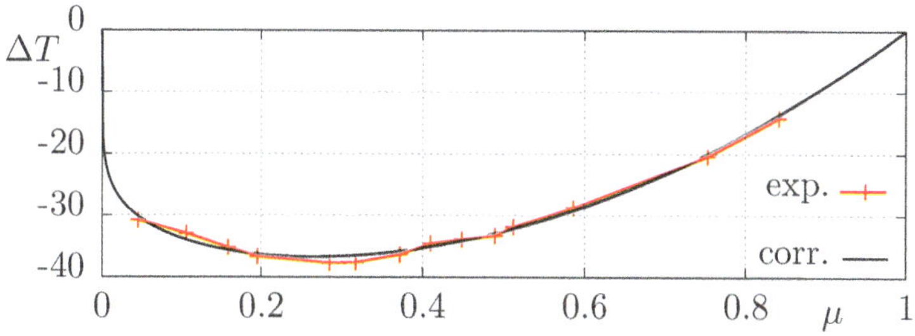

Fig. 2.3 Separation of temperature, values computed from the data Stephan et al. [21] for p_i = 5 bar and the empirical correlation (Eq. 2.9) with $n = 0.15$ and $m = 2$

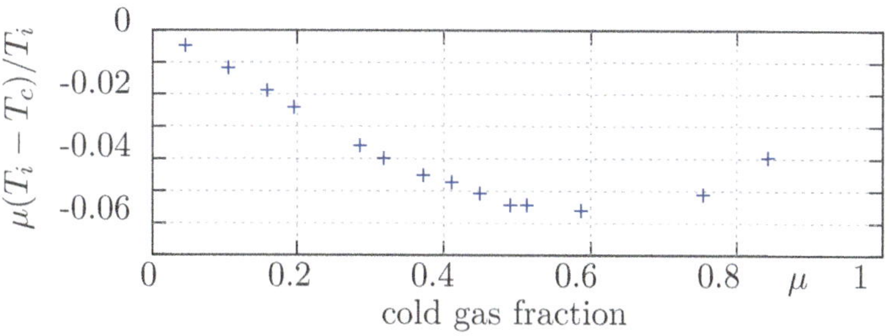

Fig. 2.4 Separation of enthalpy flux, values computed from the data Stephan et al. [21] for p_i = 5 bar

Here the factor $\left(1 + \frac{m}{n}\right)^{n/m} \left(1 + \frac{n}{m}\right)$ normalises the function $\mu^n(1 - \mu^m)$. Then ΔT has a direct physical interpretation as in the correlation presented by Stephan et al.[21]. The correlation leads to a reasonable fit to the data of Stephan et al. [21] for $p_i = 5$ bar with the parameters $n = 0.15$, $m = 2$ and $\Delta T = -36.7$ K, see Fig. 2.3. This correlation is complying with the boundary conditions at $\mu = 0$ and $\mu = 1$ as well as it is also delivering a reasonable fit to the experimental data. Experimental values and the correlation are shown in Fig. 2.3.

The separation of temperature is not identical to the separation of enthalpy flux. The enthalpy flux of the cold gas mass flux $\dot{H}_c = \dot{m}_c c_p T_c$ needs to take into account the heat capacity rate $\dot{m}_c c_p$ of the cold mass fraction. Normalising the enthalpy flux by the inlet enthalpy flux $\dot{m}_i c_p T_i$, the non-dimensional enthalpy flux separation is $\mu(T_i - T_c)/T_i$. Experimental data with air as working fluid show that the maximum enthalpy flux separation can occur at values of cold gas mass fractions of $\mu \approx 0.6$. This is illustrated by the data from Stephan [21] used to compute the specific enthalpy separation as a function of cold gas fraction μ, see Fig. 2.4.

The empirical equation of temperature difference as a function of mass fraction μ (Eq. 2.9) can be used to compute the mass fraction of maximum enthalpy separation. The corresponding empirical non-dimensional enthalpy separation differs only in the exponent of the mass fraction μ.

$$\mu \frac{T_c - T_i}{T_i} = \frac{\Delta T}{T_i} \left(1 + \frac{m}{n}\right)^{n/m} \left(1 + \frac{n}{m}\right) \mu^{n+1} (1 - \mu^m) \tag{2.10}$$

Then the value of maximum enthalpy separation occurs at

$$\mu = \left(\frac{n + 1}{m + n + 1}\right)^{1/m} \tag{2.11}$$

This establishes a direct link between the exponents m and n of the empirical equation on temperature separation and the enthalpy separation.

We summarise that from a purely thermodynamic point of view on energy conservation, it is not possible to identify the effect which transfers the energy from the cold mass fraction to the hot mass fraction.

2.3 Second Law: Entropy Balance

Application of the second law of an open system to the RHVT yields the entropy flux balance equation.

$$\dot{S}_i + \dot{S}_j + \dot{S}_c + \dot{S}_h = 0 \tag{2.12}$$

The entropy fluxes follow the thermodynamic convention of being positive when entering the control volume and negative on leaving the control volume. Compared to mass balance and energy balance an additional term due to entropy generation by friction and internal heat transfer $\dot{S}_j$ occurs. This additional term must always be larger zero $\dot{S}_j \geq 0$ since entropy cannot be decreased without heat or mass transfer.

A transfer of heat is always linked to a transfer of entropy. Therefore, the question arises if heat and entropy are transferred from the cold mass flow to the hot mass flow to establish the temperature difference. Thermodynamic principles state that heat can only spontaneously go from hot to cold and not vice versa. Therefore, entropy cannot be transferred from the cold mass fraction to the hot mass fraction. The next section will show, that from experimental data such a transfer is not necessary and the RHVT does not violate the second law.

2.3.1 Entropy Balance from Experimental Data

Since most experiments are performed with air as working fluid, the situation is examined for the case of air as an ideal, caloric perfect gas. In this case the specific entropy can be calculated.

Table 2.2 Comparison of experimental temperature difference $T_c - T_i$ to isentropic expansion temperature difference $T_{c,is} - T_i$ assuming that $T_i = 293$ K

Year	Author	p_i/p_c	$T_c - T_i$ in K	$T_{c,is} - T_i$ in K
1951	Scheper	2	−11,7	−32.6
1957	Scheller	6,1	−23	−98.2
1958	Otten	8	−50	−111.3
1968	Vennos	5,76	−13	−95.3
1969	Bruun	2	−20	−32.6
1983	Stephan	5	−38	−88.0
1983	Amitani	3,06	−19	−60.1
1988	Negm	6	−42	−97.4
1996	Ahlborn	2,7	−27	−52.4
2005	Promvonge	3,5	−30	−68.2
2005	Aljuwayhel	3	−11	−58.9

$$s(p,T) = s_0 + c_p \ln\left(\frac{T}{T_0}\right) - R\ln\left(\frac{p}{p_0}\right) \tag{2.13}$$

Assuming the air would undergo an isentropic expansion, the minimum temperature that could be achieved can be computed by the isentropic relation.

$$T_{c,is} = T_i \left(\frac{p_c}{p_i}\right)^{(\kappa-1)/\kappa} \tag{2.14}$$

Comparison of temperature difference obtained in experiments to the temperature difference obtained by ideal isentropic expansion ($T_{c,is} - T_i$) is given in Table 2.2. As the computed isentropic temperature difference is always larger than the experimentally obtained temperature difference, there is no need to transfer entropy from the cold mass flow fraction to the hot mass flow fraction to obtain a temperature separation. The isentropic temperature drop is so much larger than the experimentally observed temperature drop that there is a margin for entropy increase without contradiction to experimental findings.

2.3.2 *Analysis of Single Step Expansion*

The process of expansion on the basis of the second law is best illustrated in an $h - s$ diagram. In Fig. 2.5 the expansion from 5 bar inlet pressure to a cold outlet pressure of 1 bar and a hot outlet pressure of 1.25 bar with the measured data from Stephan et al. [21] is illustrated. Entropy was computed by Eq. 2.13 and enthalpy by equation for a caloric perfect gas $h = c_pT$. In the expansion to the cold exit and to the hot exit entropy increases. Therefore there is no violation of the second law by the expansion in the RHVT.

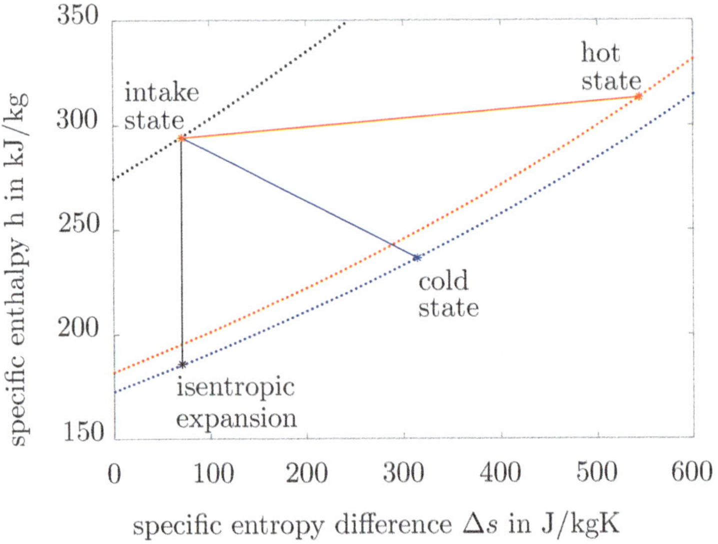

Fig. 2.5 h-s state diagram for a single step expansion with values computed from the data Stephan et al. [21] for the operation point $p_i = 5$ bar, $\mu = 0.31$, $T_{in} = 20\,°\mathrm{C}$, $T_c = -37\,°\mathrm{C}$, $T_h = 19\,°\mathrm{C}$

2.3.3 Analysis with Limited Expansion Ratio

An isentropic expansion larger than the critical expansion ratio is only possible in convergent-divergent nozzles of the de Laval type. At the inlet nozzle in the RHVT only the expansion to a critical pressure ratio is possible [27].

$$\Pi_c = \left(\frac{2}{\kappa + 1}\right)^{\kappa/(\kappa-1)} \tag{2.15}$$

The expansion in the RHVT can be regarded as a two-step process. In a first step the gas expands to the state in the nozzle with corresponding pressure and temperature. Under the assumption that the isentropic expansion attains the critical pressure ratio in the nozzle, the nozzle temperature can be computed from the isentropic relation.

$$T_n = T_i \Pi_c^{-(\kappa-1)/\kappa} \tag{2.16}$$

In a second step the gas then expands further to the cold and hot outlet states. This expansion is shown in Fig. 2.6.

It is interesting to note that total enthalpy and temperature decrease to the cold exit is larger than the isentropic expansion in the nozzle. Therefore an additional expansion with static enthalpy decrease must take place in the swirling flow of the RHVT.

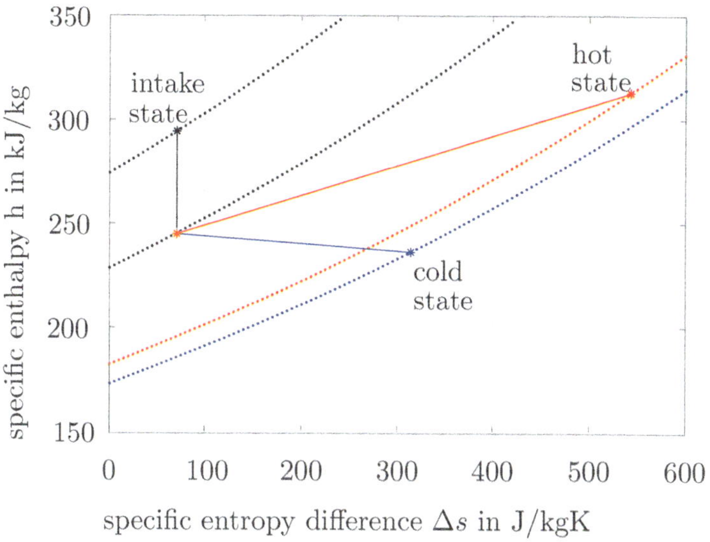

Fig. 2.6 h-s state diagram for a two-step expansion via nozzle with values computed from the data Stephan et al. [21] for the operation point $p_i = 5$ bar, $\mu = 0.31$, $T_{in} = 20\,°\text{C}$, $T_c = -37\,°\text{C}$, $T_h = 19\,°\text{C}$

2.4 Efficiency of the RHVT

If the RHVT is to be used in an application for cooling, efficiency is an important quantity as it offers information on the necessary energy to operate the device. The efficiency of a device is defined by the ratio of the benefit of the device to the effort to operate the device. This allows us different definitions: From the point of view on thermal machines, an isentropic efficiency would be the most natural choice. This would compare the enthalpy difference of the cold flow compared to the enthalpy difference generated by an isentropic expansion for the same mass flow. In the case of an ideal caloric perfect gas this yields a simple relation.

$$\eta_{is} = \frac{\dot{m}_C \Delta h_{s,C}}{\dot{m}_C \Delta h_{s,in}} = \frac{T_i - T_c}{T_i \left(1 - \left(\frac{p_C}{p_i}\right)^{(\kappa-1)/\kappa}\right)} \tag{2.17}$$

However, from an operating point of view, the effort is associated with the total mass flow and the mechanical work performed to achieve the pressure. Therefore, the cold mass fraction should enter the definition of efficiency. This is then called the coefficient of performance (COP).

$$COP = \frac{\dot{m}_C \Delta h_{s,C}}{(\dot{m}_C + \dot{m}_H)\Delta h_{s,in}} = \mu c_p \frac{T_i - T_c}{RT_i \, ln\left(\frac{p_i}{p_C}\right)} \tag{2.18}$$

The necessary mechanical work is assumed to be an isothermal compression of an ideal caloric perfect gas.

For the measured data of Stephan et al. [21] with $p_i = 5\,\text{bar}$, $\mu = 0.31$, $T_{in} = 20\,°\text{C}$, $T_c = -37\,°\text{C}$, $T_h = 19\,°\text{C}$. This leads to a COP of

$$COP = 0.31 \cdot \frac{1004\,\text{J/kgK}}{287\,\text{J/kgK}} \frac{57\,\text{K}}{293\,\text{K}\,ln\left(\frac{5}{1}\right)} = 0.13 \tag{2.19}$$

Additionally the efficiency in the process of providing the compressed air has to be taken into account here. Conventional cooling machines typically perform one order of magnitude better than the RHVT when it comes to cooling.

2.5 Thermodynamic Summary

Experimental data show no evidence of violating any thermodynamic laws. There is no need for an entropy transfer from the cold mass flux to the hot mass flux that could violate the transport equation of entropy.[1] However, there must a be a mechanism that somehow transfers energy from the cold mass fraction to the hot mass fraction. In most applications the effect of heat flux due to temperature gradient just does the opposite- it transfers energy from the hot to the cold gas. The key mechanism of transferring energy from the cold mass fraction to the hot mass fraction is more deeply analysed in the chapter on energy balance. But before going into energy balance, the flow structure of the vortex tube is to be analysed.

[1] Entropy transport is discussed in the chapter on energy transfer.

Chapter 3
Flow in the Vortex Tube

The purpose of computing is insight,
not numbers.

Richard Hamming

The flow structure in the vortex tube can easily be simulated by CFD tools. A sketch of what is believed to be the basic flow structure for a counter-flow RHVT is given in Fig. 3.1. A streamline of hot air -illustrated by the red line- starts from the tangential inlet swirling through the pipe next to the wall to reach the annular outlet gap at the outer tube radius. The blue line is a streamline of air leaving the RHVT as cold air simultaneously starts at the tangential inlet and swirls to the cold exit while constantly remaining at a smaller radius than the hot air streamline. This is an important detail to be remembered when we come to the energy balance equation in the next chapter.

At the inlet the fluids of both streamlines have the same initial conditions, identical temperature and velocities. Velocity in the circumferential direction and positive axial direction are largest in a small layer at the outer wall. The high velocities cause a strong radial velocity gradient next to the wall. The flow in the centre of the RHVT leaves the tube at the cold outlet. In the case of the counter-flow RHVT, the axial flow in the centre is in the opposite direction of the axial flow at the wall. This causes a more dominant axial velocity gradient over the tube radius than there is in the case of the co-flow RHVT.

Both streamlines follow a path close to the cylinder wall. To a certain degree their circumferential velocities are similar to the circumferential velocities of a forced vortex, whereas the axial velocities are dominant when at the cylinder wall interface.

Electronic Supplementary Material The online version of this chapter (https://doi.org/10.1007/978-3-030-89766-6_3) contains supplementary material, which is available to authorized users.

A, Kaufmann, *The Ranque Hilsch Vortex Tube Demystified*,
https://doi.org/10.1007/978-3-030-89766-6_3

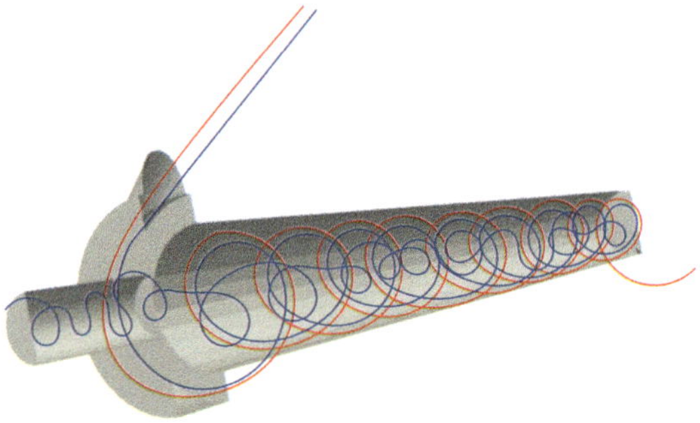

Fig. 3.1 Sketch of the flow in the counter-flow RHVT

This is essentially due to the forcing of the tangential velocity by the inlet velocity boundary condition.

This flow structure was found in many CFD computations ranging from Reynolds Averaged Navier Stokes (RANS) computations to Large Eddy Simulations (LES) [20]. However, LES tend to show results closer to the experimental values [4, 10, 11] with respect to energy separation.

For the understanding of the flow structure it is useful to start off with the limiting cases of the mass flow fraction.

In the case of $\mu = 0$ all of the flow into the vortex tube is leaving the vortex tube at the circumferential exit at the hot end. Since the total enthalpy is conserved as discussed in the previous chapter, the total enthalpy of the flow leaving the tube has the identical total temperature as the inlet. The swirling velocity will be reduced due to the friction on the cylinder wall. This flow structure can be mimicked by a tube with a solid body rotation at the inlet.

In the other case with a mass flow coefficient $\mu = 1$ all of the flow entering the vortex tube leaves the tube at the cold orifice. Again the total enthalpy is conserved and leaves the cold orifice with the same total enthalpy as at the inlet. In this case the angular momentum balance requires an increase in circumferential velocity. The flow structure can be represented by the flow structure of a bidirectional vortex as described by Vyas et al. [24].

In the case of the RHVT, two outlets are present and for the balance of the angular momentum it is not necessary to form a free vortex type flow if angular momentum is transferred. In the case of $\mu < 1$ some momentum can be transferred to the flow leaving the hot outlet.

This chapter starts with the origin of the flow structure in the vortex tube for the case of $\mu = 0$ when all the flow is leaving at the hot exit. The flow structure is best understood by starting with a simplified geometry: In this case a pipe with a swirling flow inside caused by a forced vortex inlet is chosen. This differs from the boundary condition imposed by the tangential inlet of the vortex tube but allows in a simplified setup to understand the impact of the degree of swirl on the flow structure in the tube. It is shown that for low circumferential velocities compared to the axial velocity, the flow does not change much compared to the solid body rotation. In the case of high circumferential velocities compared to the axial velocity, the flow structure changes substantially. The example of swirling flow in the pipe will be referred to in the chapter on energy balance.

In the second step CFD results of the flow in a RHVT geometry by Farouk [10, 11] are reproduced and analysed for the limiting cases of $\mu = 0$ and $\mu = 1$ as well as the case of maximum temperature separation $\mu = 0.3$.

3.1 Structure of Swirling Flow with a Forced Vortex Boundary Condition

The structure of the swirling flow in a pipe depends on a great extent on the boundary conditions. Here we consider a particularly simple example with a solid body rotation as inlet boundaries. To some extent those are similar to the boundary conditions used in the case of the RHVT.

To illustrate the flow structure, results of incompressible LES of the tube are shown. The geometry consists of a cylinder with a diameter $d = 2\,r_0$ and a length $l \approx 6d$. On the inlet side of the cylinder a boundary condition with a flow in the axial direction u_z is imposed. Superimposed is a tangential velocity u_θ resulting in a velocity profile of a solid body rotation at the inlet (forced vortex). The unsteady simulation is carried out for roughly 10 convective times with respect to the mean axial flow velocity. The last 5 convective times are time-averaged to obtain the average flow field.

In Fig. 3.2 the inlet boundary condition with $u_\theta = u_{\theta 0} \cdot r/r_0$ is illustrated for the case in which the outermost tangential velocity $u_{\theta 0}$ is 20 times the axial velocity u_z ($u_{\theta 0}/u_{z0} = 20$). The size of the arrow reflects the magnitude and direction of the velocity.

Figure 3.3 shows a cut through the tube with time-averaged pressure and velocities in the cylinder coordinate system (axis z, radial r and θ angle) for a boundary velocity ratio of $u_{\theta 0}/u_{z0} = 20$.

Even if the axial inlet velocity is uniform $u_{z0} = 10\,\text{m/s}$, the axial velocity increases just inside the cylinder wall to 50 m/s showing a strong radial gradient

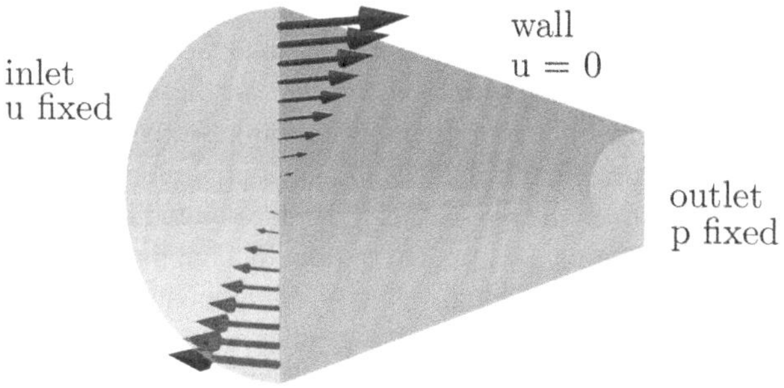

Fig. 3.2 Boundary condition for the turbulent pipe

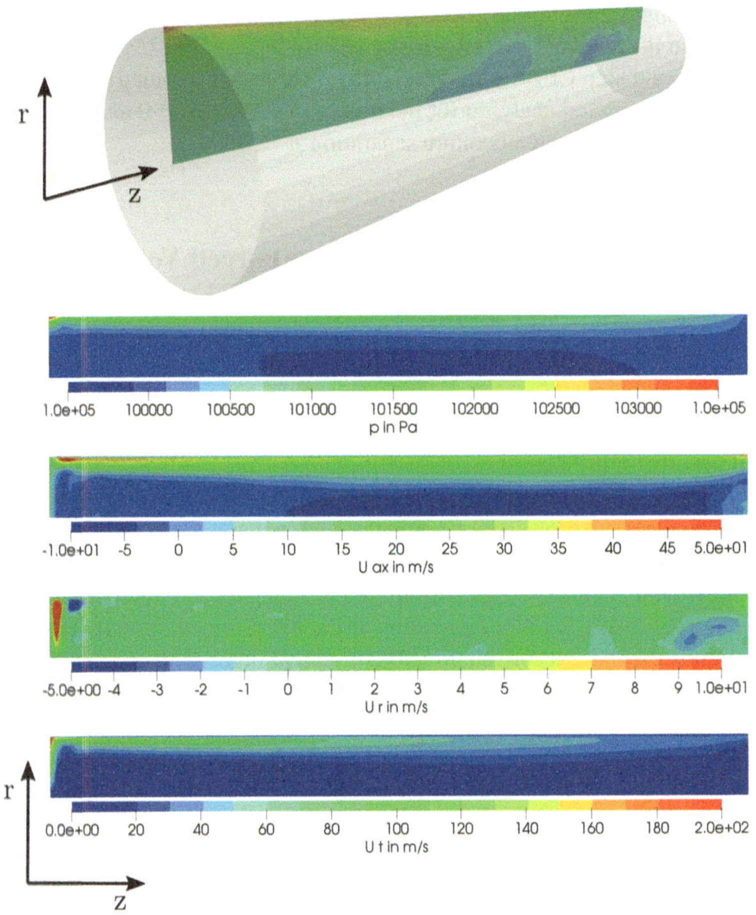

Fig. 3.3 Cross section of the pipe. The figure from top to bottom shows mean pressure, axial velocity, radial velocity magnitude and tangential velocity for the case $u_{\theta 0}/u_{z0} = 20$

of the axial velocity from the axis to the wall. The tangential velocity u_θ shows a strong radial gradient, too. It increases from 0 to almost 200 m/s just inside the cylinder wall as imposed by the forced vortex boundary condition.

Most interesting is the radial velocity profile u_r next to the inlet boundary condition. Next to the inlet boundary there is a motion from the axis to the cylinder wall. This has two effects: First it reduces the tangential momentum on the outside since the flow coming from the axis needs to be accelerated in tangential direction, and second, it is also creating a recirculating flow in axial direction.

Next, the effect of the tangential velocity component is to be illustrated. In Fig. 3.4 the normalised averaged velocity profiles at the inlet $z/d = 0$ and at $z/d = 2.5$ are shown for different magnitudes of swirl ($u_{\theta 0}/u_{z0} = 1$, $u_{\theta 0}/u_{z0} = 5$ and $u_{\theta 0}/u_{z0} = 20$). In the case of smaller tangential velocities $u_{\theta 0}/u_{z0} = 1$ the flow

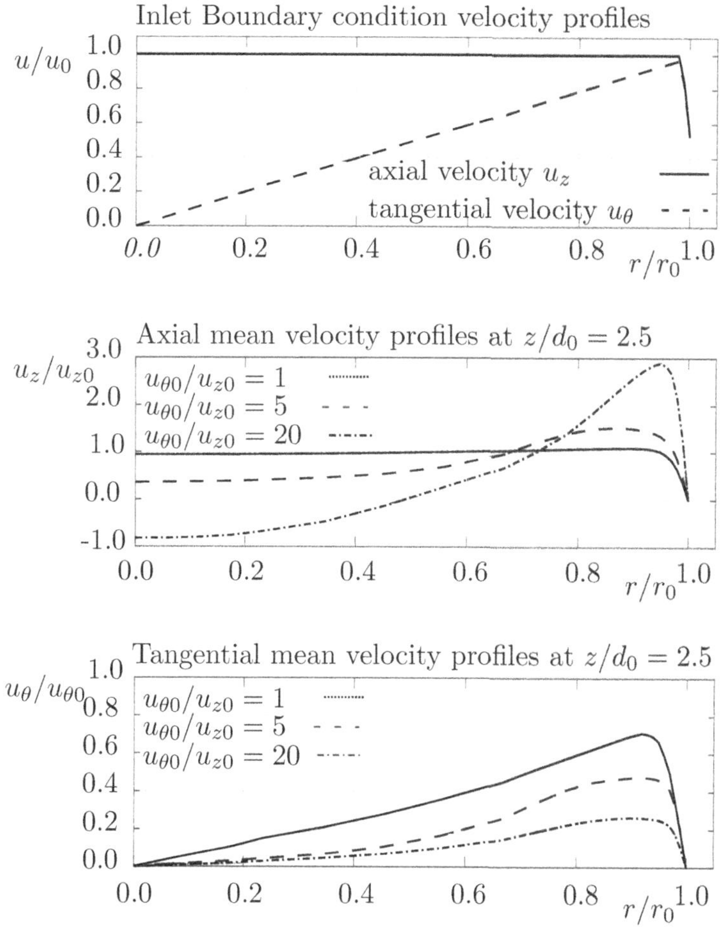

Fig. 3.4 Normalised mean velocities in a turbulent pipe with solid body rotation as inlet boundary, normalisation by inlet boundary conditions

maintains a structure of the solid body rotation. The boundary layer on the cylinder wall throughout the tube does not differ from the flow in a pipe. The flow profiles are marked in Fig. 3.4 with a dotted line.

Increasing the tangential velocity at the boundary to $u_{\theta 0}/u_{z0} = 5$ shows a change in the axial and tangential velocity profiles (dashed line). Axial and tangential velocities decrease in the centre of the tube and increase to a peak before the boundary layer. The tangential velocity u_θ starts to differ from a solid body rotation. A further increase in the tangential velocity at the boundary to $u_{\theta 0}/u_{z0} = 20$ even results in negative axial velocities u_z. The impact of the circumferential velocity on the axial velocity is not intuitive as one would expect a constant solid body rotation throughout the pipe. In the next section, the Euler equations for the flow are used to understand the origin of this flow structure.

3.1.1 Secondary Flow Structures in the Turbulent Pipe

Velocity profiles and secondary flow structures can be explained by the steady Euler-equations for incompressible flow

$$\begin{aligned} \nabla.\vec{u} &= 0 \\ \rho\vec{u}.\nabla\vec{u} &= -\nabla p \end{aligned} \tag{3.1}$$

For the sake of simplification, the equations are used in cylindrical coordinates and the dependence on the angular variable θ is ignored. Although this results in a set of equations for continuity and momentum, more simplifications are desirable for an analytical solution.

$$\begin{aligned} \frac{1}{r}\frac{\partial}{\partial r}ru_r + \frac{\partial}{\partial z}u_z &= 0 \\ u_r\frac{\partial}{\partial r}u_r + u_z\frac{\partial}{\partial z}u_r - \frac{u_\theta^2}{r} &= -\frac{1}{\rho}\frac{\partial}{\partial r}p \\ u_r\frac{\partial}{\partial r}u_\theta + u_z\frac{\partial}{\partial z}u_\theta + \frac{u_\theta u_r}{r} &= 0 \\ u_r\frac{\partial}{\partial r}u_z + u_z\frac{\partial}{\partial z}u_z &= -\frac{1}{\rho}\frac{\partial}{\partial z}p \end{aligned} \tag{3.2}$$

In most axisymmetric solutions, the momentum equation for u_θ is simplified to

$$u_r\frac{\partial}{\partial r}u_\theta + \frac{u_\theta u_r}{r} = u_r\left(\frac{\partial}{\partial r}u_\theta + \frac{u_\theta}{r}\right) = 0 \tag{3.3}$$

by ignoring the change of tangential velocity u_θ in the axial direction $u_z\frac{\partial}{\partial z}u_\theta$. Two principle solutions exist. One solution satisfying those conditions is $u_r = 0$. The other solution of this equation is the tangential velocity of a free vortex.

$$u_\theta(r) = \frac{c}{r} \tag{3.4}$$

But both solutions evidently contradict the results of the LES computations and the previous assumption does not hold true. This makes it necessary to take the change of $\partial u_\theta/\partial z$ along the axis into account. Next to the inlet boundary condition, the tangential velocity profile is approximately a solid body rotation and decreases

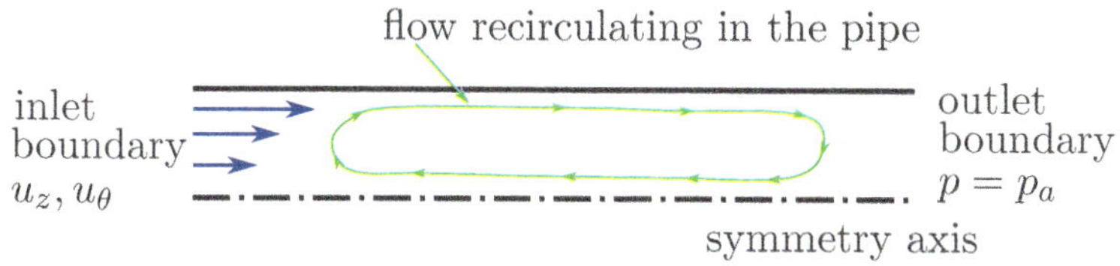

Fig. 3.5 Flow structure in the turbulent pipe with a solid body rotation as inlet boundary condition

due to friction in the direction of the z axis. For the analysis, the velocity profile is assumed to split into a radial function and an axial function next to the inlet boundary condition.

$$u_\theta = u_{\theta,0}\left(\frac{r}{r_0}\right) g(z) \tag{3.5}$$

The u_θ velocity can be introduced into the corresponding Euler momentum equation.

$$u_r \frac{\partial}{\partial r} u_\theta + u_z \frac{\partial}{\partial z} u_\theta + \frac{u_\theta u_r}{r} = u_r \left(\frac{\partial}{\partial r} u_\theta + \frac{u_\theta}{r}\right) + u_z \frac{\partial}{\partial z} u_\theta = 0 \tag{3.6}$$

This brings about an expression for the radial velocity next to the inlet boundary condition by solving for u_r in

$$u_r \left(\frac{u_{\theta,0}}{r_0} + \frac{u_{\theta,0}}{r_0}\right) g(z) + u_z u_{\theta,0}\left(\frac{r}{r_0}\right) \frac{\partial g(z)}{\partial z} = 0 \tag{3.7}$$

$$u_r = -\frac{u_z r}{2g(z)} \frac{\partial g(z)}{\partial z} \tag{3.8}$$

since the tangential velocity u_θ decreases along the axis and the gradient is negative $\frac{\partial g(z)}{\partial z} < 0$. This produces a positive u_r velocity next to the inlet boundary. Continuity requires the flow in the axial direction u_z to accelerate next to the cylinder wall and to decrease in the centre of the tube. The structure of the flow is sketched in Fig. 3.5. Since the decay gradient $\frac{\partial g(z)}{\partial z}$ is stronger for increasing $u_{\theta,0}$, the velocity u_z becomes faster next to the cylinder wall. This explains the shape of the velocity profiles in Fig. 3.4.

Any reader unfamiliar with the handling of differential equations should simply recall that the boundary conditions at the inlet have a strong impact on the flow structure. The flow profile inside the tube is not necessarily identical to the flow profile imposed on the inlet boundary and a secondary flow structure can arise.

In the case of the RHVT the inlet boundary conditions are imposed at the outer circumference of the cylinder. The resulting velocity profile differs from the solid body rotation (forced vortex), but still there is a strong positive radial gradient present. This produces similar effects and creates a circulation in axial direction. This will be illustrated when the flow structure of the RHVT is presented.

3.1.2 Secondary Flow Structures at the Cylinder Wall

The flow in the turbulent pipe is certainly unsteady and shows numerous small scale structures. Depending on the intensity of the rotational component u_θ different structures appear and contribute to an enhanced mixing of the flow.

High resolution LES of the turbulent flow with a solid vortex of high circumferential speed show numerous small scale structures on the tube wall. On those curved walls, Görtler vortices may appear under certain conditions. Stephan et al. [21] argue that Görtler vortices play an important role in the energy separation but do not provide a precise argument.

3.2 Flow Structure and Typical Velocity Profiles in the RHVT

In the previous section the simplified geometry of the turbulent pipe with a solid body rotation as inlet boundary condition was used to introduce the basic flow structure. In this section a more realistic geometry with realistic boundary conditions is analysed. Farouk et al. [10, 11] published on LES computations with detailed velocity profiles and temperature fields. The geometry used by Farouk is representative of a typical RHVT geometry (see Fig. 3.6). It can be considered representative to illustrate the possible flow structure in a counter-flow RHVT. Further it serves as an example to analyse the viscous source terms in the chapter on energy balance.

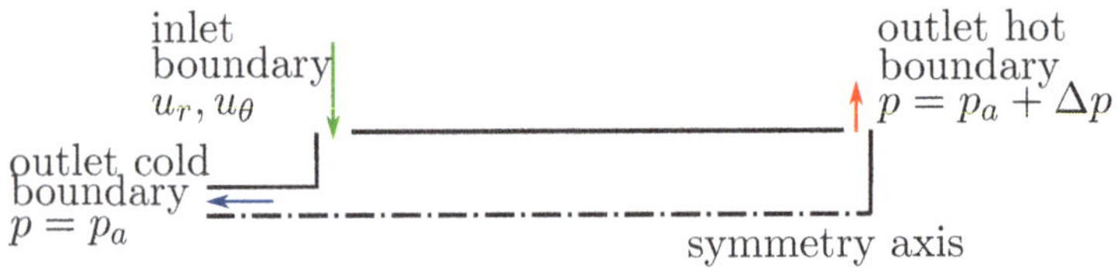

Fig. 3.6 Sketch of the geometry used by Farouk et al.

Geometry of the Farouk Tube

The inner tube has a length of $l = 100\,\text{mm}$ and a diameter of $d = 20\,\text{mm}$. Inlet boundary conditions are imposed on a circumferential ring of $b = 3\,\text{mm}$ width. The hot outlet boundary is a circumferential ring of $c = 1.5\,\text{mm}$ width and the cold outlet tube has a diameter of 6 mm.

Boundary Conditions for the Simulation

Most publications on CFD in the RHVT present parametric studies on the boundary conditions to study the impact of different mass flow ratio μ. Here, the presentation is limited to one single combination of boundary conditions that leads to a mass flow ratio of $\mu = 0.3$. This leads to a tangential inlet velocity of $u_\theta = 300\,\text{m/s}$ and a radial inlet velocity of $u_r = -10\,\text{m/s}$. On the hot outlet a constant ambient pressure of $p_a = 1\,bar$ is imposed. The mass flow splitting is achieved by imposing the corresponding volume flow rate on the cold outlet section.

Justification of Incompressible Analysis

Let us suppose the flow has a solid body rotation tangential flow profile $u_\theta = u_{\theta,max}\left(\frac{r}{r_0}\right)$. Based on the Euler equations the pressure difference from the centre exit to the inlet can be computed as follows:

$$\frac{u_\theta^2}{r} = \frac{1}{\rho}\frac{\partial p}{\partial r} \tag{3.9}$$

For incompressible flow with a density of the order of unity $\rho = 1\,\text{kg/m}^3$ and a tangential velocity in the order of 200 m/s this leads to a pressure difference of

$$\Delta p = \frac{1}{2}\rho u_{\theta,max}^2 = 2 \cdot 10^4\ Pa \tag{3.10}$$

If the exit pressure (ambient) is 1 *bar* this corresponds to 20% of the total pressure and is therefore acceptable for an incompressible flow computation. In the case of a flow coefficient of $\mu = 0$ this estimation is valid. In the case of $\mu = 1$, however, the flow has the structure of a bidirectional vortex and the pressure difference is much larger. The criterium for incompressible flow is not any more fulfilled. However, the flow structure is not expected to change significantly and for the sake of simplification, the computations are performed with constant density.

Estimation of Boundary Layer and Grid Spacing

The estimation of the boundary layer thickness is based on a velocity of $u_0 = 200\,\text{m/s}$ and a characteristic length scale of $x = 100\,\text{mm}$ with properties of ambient air. Using the standard approximation, see Schlichting [18], this leads to the following turbulent boundary layer thickness δ.

$$\delta \approx 0,37\frac{x}{Re^{1/5}} = 0,37x^{4/5}\left(\frac{\nu}{u_{max}}\right)^{1/5} = 3,5 \cdot 10^{-4}\,\text{m} \tag{3.11}$$

The LES computation requires $y^+ = 1$ for the boundary grid spacing. The corresponding grid spacing Δs is computed using the friction coefficient proposed by Schlichting.

$$
\begin{aligned}
C_f &= \left(2\log_{10}(Re) - 0,65\right)^{-2,3} = 0,00479 \\
\frac{\tau_w}{\rho} &= C_F \frac{1}{2} u_{max}^2 = 95,7\,\mathrm{m^2/s^2} \\
u^* &= \sqrt{\frac{\tau_w}{\rho}} = \sqrt{C_F \frac{1}{2} u_{max}^2} = 9,78\,\mathrm{m/s} \\
y_W &= y^+ \frac{\nu}{u^*} = 1,53 \cdot 10^{-6}\,\mathrm{m}
\end{aligned}
$$

This leads to a necessary grid spacing of $\Delta s = 1,5 \cdot 10^{-6}$ m at the wall and a time step limitation (CFL criterium) of roughly $\Delta t < 2 \cdot 10^{-9}$ s.

The near wall grid spacing of $\Delta s = 1,5 \cdot 10^{-6}$ m requires an adaptive mesh, since a uniform mesh would lead to a prohibitive number of grid cells. To achieve the best possible numerical result, a hexa-mesh with a refinement next to the wall is used. Figure 3.7 shows a cut of the grid geometry. The refinement for the boundary layer is shown in the lower part of the figure.

Justification of the Continuum Hypothesis

The small magnitude of the boundary layer thickness and the high velocities lead to the question if the continuum hypothesis is still adequate for the analysis based on Navier–Stokes equations. This can be done by evaluating the Knudsen number. The Knudsen number compares the mean free path length (λ) with the physical length scale (L). For gases the Kundsen number can be obtained from Mach number (Ma) and Reynolds number (Re) in the equation.

$$
Kn = \frac{\lambda}{L} = \frac{Ma}{Re}\sqrt{\frac{\kappa\pi}{2}} \tag{3.12}
$$

To obtain the maximal Knudsen number in the RHVT, we employ the assumption that the Mach number attains unity $Ma = 1$ and take a Reynolds number based on a length scale smaller than the thickness of the boundary layer $\delta = 1 \cdot 10^{-4}$ m.

$$
Re = \frac{300\,\mathrm{m/s} \cdot 10^{-4}\,\mathrm{m}}{1.5 \cdot 10^{-5}\,\mathrm{m^2/s}} = 2 \cdot 10^{-3} \tag{3.13}
$$

$$
Kn = \frac{1}{2 \cdot 10^3}\sqrt{\frac{1.4 \cdot \pi}{2}} = 0.00074 \tag{3.14}
$$

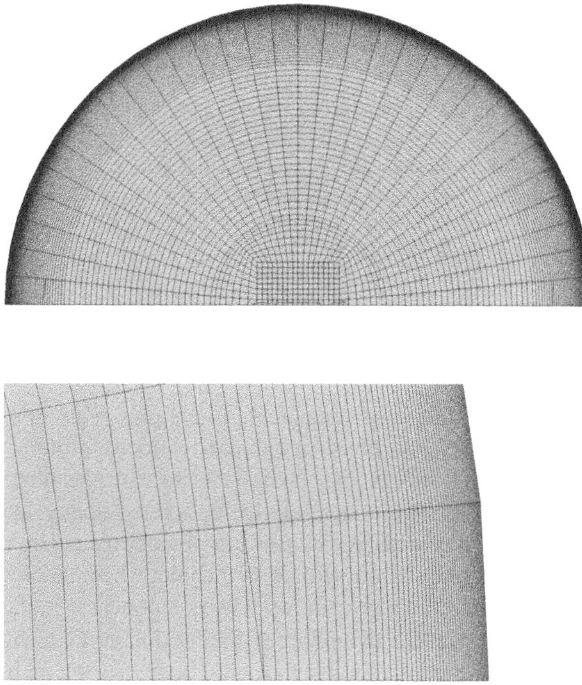

Fig. 3.7 Cut of the geometry with grid, top half of the tube and boundary layer resolution

This is still one order of magnitude smaller than the limiting Knudsen number of Kn = 0.01. This proves that the continuum hypothesis holds true and CFD computations can be expected to deliver some information on the structure of the flow.

The CFD computation is performed using a modified version of the *pisoFoam* solver from the *OpenFoam* [15] package.

3.2.1 Sections for the Velocity Profiles

To give more than colourful pictures of the magnitude of the flow, some sections for the presentation of velocity profiles are defined. The locations of the cut lines are shown in Fig. 3.8.

In the following sections the flow structure obtained by Large Eddy Simulation for the three cases of flow coefficient is described. The results will be helpful in the next chapter when reflecting on the energy transfer in the vortex tube.

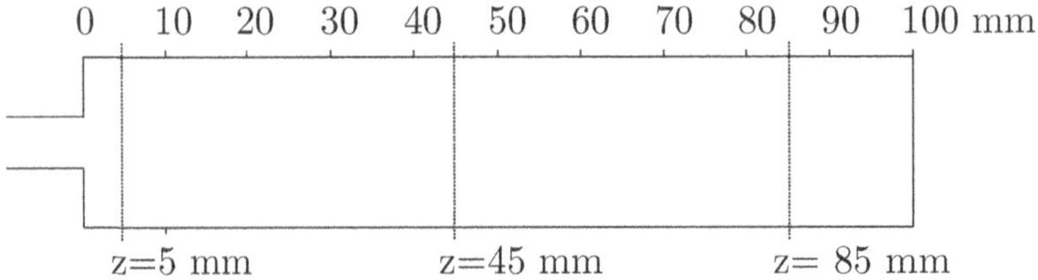

Fig. 3.8 Cut sections for velocity profiles

3.2.2 *RHVT Flow Structure with* $\mu = 0.0$

In this case the flow has structures comparable to the pipe with a forced vortex as inlet condition. Streamlines and the time-averaged velocity profiles are shown in Fig. 3.9. There is a high axial velocity component next to the tube wall into the direction of the hot outlet. The magnitude of the axial velocity lowers with the increase of the boundary layer thickness. Towards smaller diameters the velocity is reversed resulting in a volume flow rate back to the inlet where the flow is entrained again to the hot exit. The circumferential velocity component is highest next to the tube wall and decreased towards the outlet by the friction at the wall. In the centre of tube the velocity profile is comparable to a solid body rotation with low friction.

3.2.3 *RHVT Flow Structure with* $\mu = 1.0$

In this case the flow structure is comparable to the flow structure of the bidirectional vortex. Streamlines and time-averaged velocity profiles are shown in Fig. 3.10. The flow enters the tube with a circumferential velocity and proceeds along the tube wall with a high axial velocity component to the opposed end of the tube. There the streamlines indicate that the flow is reversed at a diameter smaller than the cold outlet orifice. The balance of angular momentum requires the circumferential velocity to increase towards the centre of the tube. At diameters down from roughly 1/2 of the outlet orifice, the circumferential velocity profile is comparable to a solid body rotation. Since the angular momentum balance does not comply with a steady state solution, the flow is very unstable in the centre of the tube. Due to the increase of circumferential velocity towards the centre of the tube, the pressure difference from the inlet to the outlet of the tube is significantly higher than in the case of $\mu = 0$.

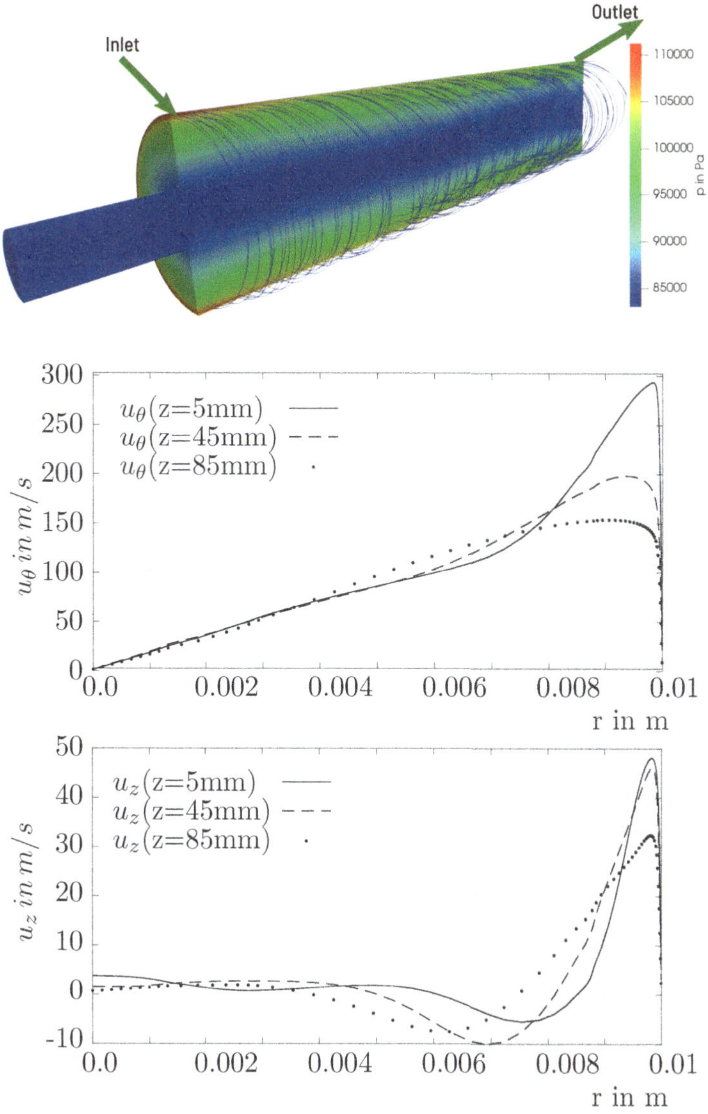

Fig. 3.9 Top: streamlines computed from the time-averaged velocity for the case $\mu = 0$ starting from the inlet, Below: time-averaged circumferential and axial velocity profiles

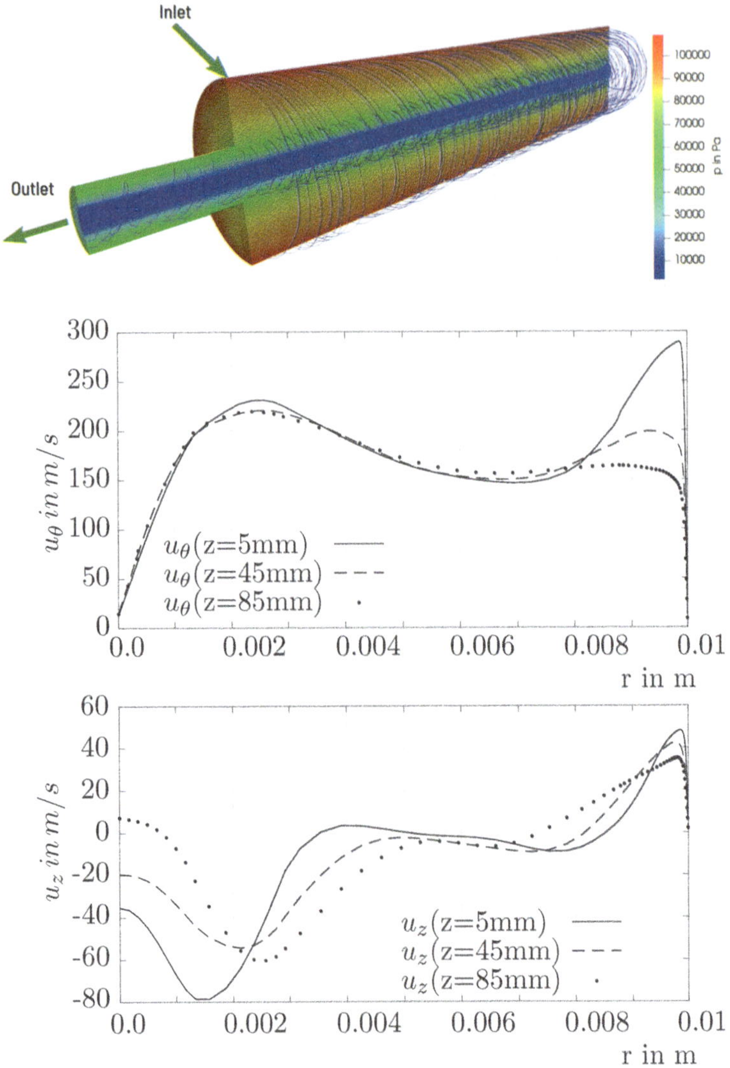

Fig. 3.10 Top: streamlines computed from the time-averaged velocity for the case $\mu = 1$ starting from the inlet, Below: time-averaged circumferential and axial velocity profiles

3.2.4 Flow Structure at $\mu = 0.3$

This case will show features of the flow structures from the two previous cases. A simplified point of view is to superpose the flow structures of the cases of $\mu = 0$ and $\mu = 1$ to get the flow structure of the other cases. Figure 3.11 shows the time-averaged flow fields. The radial pressure difference is also in between the pressure ratios of the previously discussed cases and exceeds 50%. This is due to the circumferential velocity structure shown in the middle figure. High velocities, both in axial and circumferential direction, occur next to the cylinder walls. In the centre velocities are significantly smaller. The strong axial velocities near the boundary compared to the centre of the vortex tube determine the volume flow in axial direction to be predominantly in a thin layer of the order of 1 mm next to the wall. Quantitatively this is shown in Fig. 3.14.

In Fig. 3.12 some instantaneous flow fields are shown. Numerous small scale structures appear in the flow field. Those are best made visible by iso-surfaces of the Q-criterion, color-coded by the magnitude of velocity as shown in Fig. 3.13. The flow in the vortex tube is turbulent and does by no means have as simple a structure as the streamlines in the introduction would have suggested. This is best illustrated by a transient simulation.

Movies of transient simulations can be found in the supplementary material.

The time-averaged velocity profiles in Fig. 3.14 reveal very strong velocity gradients in the boundary layer next to the wall. The boundary layer thickness seems to vary from roughly $\delta \approx 1.2 \cdot 10^{-4}$ m next to the inlet to roughly $\delta \approx 5 \cdot 10^{-4}$ m next to the outlet. This corresponds to the boundary layer thickness previously estimated by the Reynolds number.

The difference in the velocity profile to a solid body rotation or a free vortex in the total cross section is evident. In the center of the tube, from the axis to roughly the radius of the cold outlet, the flow has the structure of a forced vortex. For larger radii up to roughly 3/4 of the tube radius it rather has the structure of a free vortex. Right next to the wall the circumferential velocity is only largest next to the inlet and controls the flow rate from inlet to outlet.

The information from the velocity profiles can be assembled to a sketch of the average axial flow velocity as shown in Fig. 3.15. Here it is helpful to calculate the local volume flow rates $\Delta \dot{V}_z(r) = u_z \cdot 2\pi r \cdot dr$ and $\Delta \dot{V}_r(r) = u_r \cdot 2\pi r \cdot dr$. The normalised volume flow rates are given in Fig. 3.15 too. The flow enters the tube through the nozzle tangential to the wall at very high velocity. It proceeds in axial direction along the tube wall at high velocity. At the other end of the tube the flow is split. One fraction leaves at the hot exit while the other fraction is guided by the rear wall to a smaller radius. This fraction of the flow is swirling back to the cold outlet. The dominant flow fraction in axial direction occurs roughly at the radius of the cold outlet.

In order to estimate the amount of recirculation, the volume flow rate with positive contributions near the tube wall is integrated over the flow cross section. Compared to the inlet flow rate, this leads to an access of 19% at the section z=5 mm,

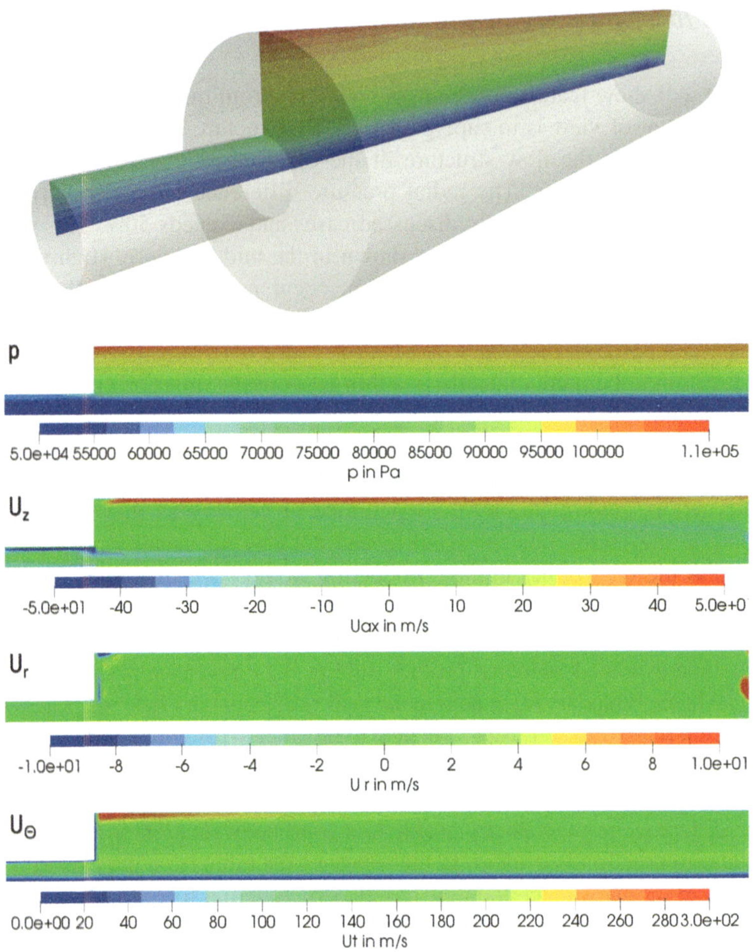

Fig. 3.11 Cross section of the geometry, time-averaged pressure and velocities, $\mu = 0.3$

45% at the section z=45 mm and 46% at the section z=85 mm. This illustrates that there is a significant recirculation occurring next to the tube wall. Those flow profiles have an important impact on the enthalpy balance which will be discussed in the following chapter.

3.3 Circumferential Velocity Balance

Taking a detailed look at the circumferential velocity reveals the necessity for the flow to be instationary. This becomes clear by taking the stationary Reynolds averaged circumferential velocity equation in cylinder coordinates.

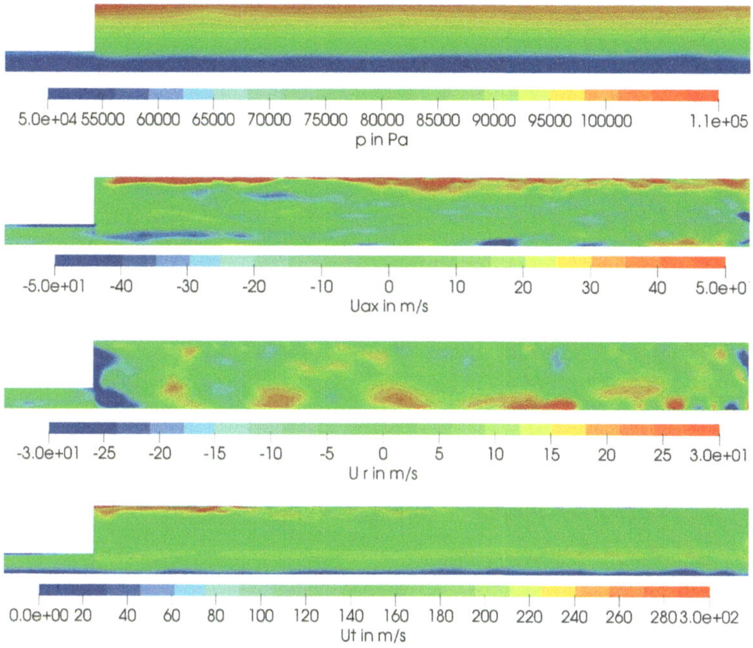

Fig. 3.12 Cut of the geometry along the axis, instantaneous pressure and velocities, $\mu = 0.3$

Fig. 3.13 Iso-Q surfaces color-coded by the velocity show the numerous vortex structures in the RHVT, $\mu = 0.3$

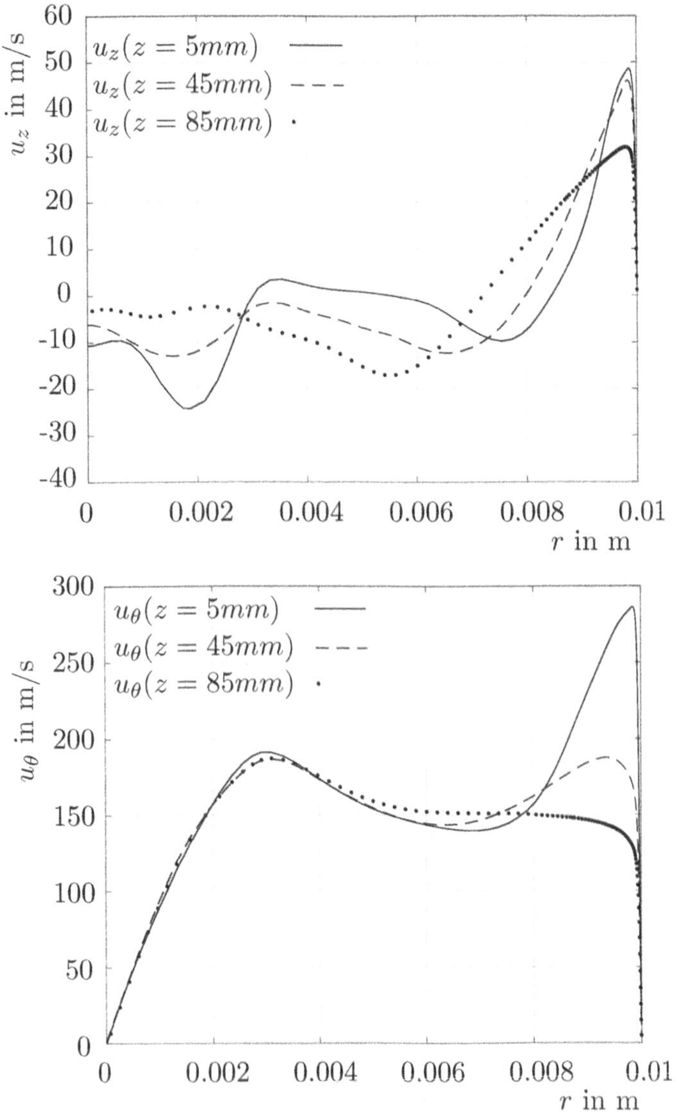

Fig. 3.14 Time-averaged axial and circumferential velocities at different sections

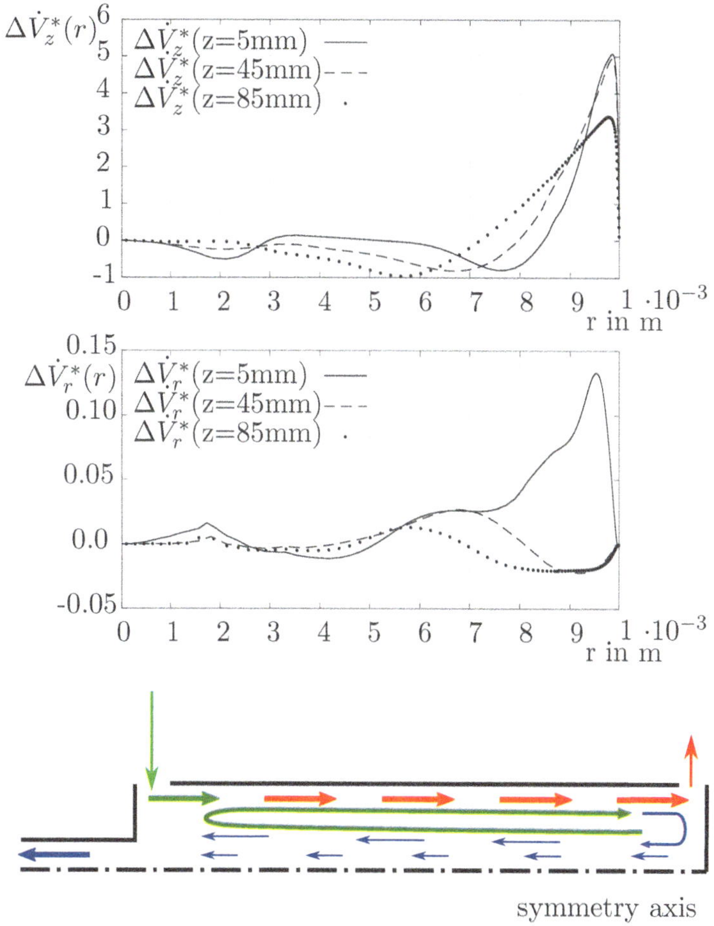

Fig. 3.15 Volumetric flow rates and a sketch of the time-averaged axial flow structure in the RHVT

$$
\begin{aligned}
\rho\left(\bar{u}_r\frac{\partial}{\partial r}\bar{u}_\theta+\frac{\bar{u}_\theta}{r}\frac{\partial}{\partial\theta}\bar{u}_\theta+\bar{u}_z\frac{\partial}{\partial z}\bar{u}_\theta+\frac{\bar{u}_r\bar{u}_\theta}{r}\right)&=-\frac{1}{r}\frac{\partial p}{\partial\theta}\\
+\mu\left(\frac{1}{r}\frac{\partial}{\partial r}\left(r\frac{\partial\bar{u}_\theta}{\partial r}\right)+\frac{1}{r^2}\frac{\partial^2}{\partial\theta^2}\bar{u}_\theta+\frac{\partial^2}{\partial z^2}\bar{u}_\theta+\frac{2}{r^2}\frac{\partial}{\partial\theta}\bar{u}_r-\frac{\bar{u}_\theta}{r^2}\right)&\\
-\overline{u'_r\frac{\partial u'_\theta}{\partial r}}-\overline{\frac{u'_\theta}{r}\frac{\partial u'_\theta}{\partial\theta}}-\overline{u'_z\frac{\partial u'_\theta}{\partial z}}-\overline{\frac{u'_r u'_\theta}{r}}&
\end{aligned}
\tag{3.15}
$$

With exception to the wall, the averaged velocity gradients are too small for the viscosity term to generate a significant contribution to the balance equation. Assuming an axisymmetric Reynolds averaged solution, the dominant terms in the averaged transport equation are obtained by keeping the circumferential velocity components.

$$\bar{u}_r \left(\frac{\partial \bar{u}_\theta}{\partial r} + \frac{\bar{u}_\theta}{r} \right) = 0 \quad (3.16)$$

Since the radial velocity $\bar{u}_r$ cannot be zero in regions where the circumferential momentum is transported to the center of the tube, the second term needs to vanish. This has the previously discussed solution with increasing circumferential velocity at the center of the tube.

$$\bar{u}_\theta(r) = \bar{u}_{\theta,r0} \frac{r_0}{r} \quad (3.17)$$

This solution is incompatible with the axisymmetric Reynolds averaged solution $\bar{u}_\theta(r = 0) = 0$. The balancing of the circumferential momentum can occur by the axial transport $\bar{u}_z \frac{\partial}{\partial z} \bar{u}_\theta$ or the fluctuating contributions in the last line of Eq. 3.15. Since the axial transport is not visible in the equation to a magnitude that could explain the circumferential velocity balance (see Fig. 3.14), at a radius corresponding to the cold outlet, the flow must be unsteady. This might be the origin of the strong noise generation in the operation of the vortex tube.

3.4 Summary of the Flow Structure in the Vortex Tube

The boundary conditions with the high circumferential velocities lead to a complex flow structure in the tube. This structure cannot be simply modelled with a free or a forced vortex. High circumferential and axial velocities occur next to the tube wall, leading to a highly turbulent boundary layer with a strong radial pressure gradient. The high axial velocities in the near wall region dominate the flow rate from the inlet to the hot outlet.

Still the principal shape of the velocity profiles can be used to investigate in the impact on the enthalpy balance in the RHVT. This will be the subject of the next chapter.

Chapter 4
The Energy Transport Mechanism

> The law of conservation of energy tells us we can't get something for nothing, but we refuse to believe it.
>
> Isaac Asimov

The remaining question from Chap. 2 on thermodynamics of the RHVT is about the working principle of energy separation in the RHVT.

Here it should be recalled that in a first step the reduction of static temperature results from the adiabatic expansion from the reservoir state to the high velocity state in the nozzle. This is not only obvious for the case of the laval nozzle but also holds true for the swirling flow since the internal energy is transferred to a correlated velocity. When heat transfer is not considered, the total enthalpy must be conserved. Therefore in this chapter the balance equation for the total enthalpy is examined in detail. The total enthalpy includes the kinetic energy, and therefore the velocity profiles from the previous chapter play an important role. The shape of the velocity profiles is shown to play an important role on the viscous term that is essential in the energy balance equation.

4.1 Energy Conservation Equations

Conservation equations for the different energy equations can be found in standard text books on fluid dynamics [2, 5, 6] or turbulence [7]. Here the symbol convention found in the books of continuum mechanics [1] is used. We will begin with the

Electronic Supplementary Material The online version of this chapter (https://doi.org/10.1007/978-3-030-89766-6_4) contains supplementary material, which is available to authorized users.

A, Kaufmann, *The Ranque Hilsch Vortex Tube Demystified*,
https://doi.org/10.1007/978-3-030-89766-6_4

conservation of total enthalpy and work our way to kinetic energy and sensible enthalpy.

4.1.1 Conservation of Total Enthalpy

In the chapter on the thermodynamic point of view, the importance of the enthalpy balance 2.3 was stressed. The corresponding field equation for the total enthalpy $h_t = h_s + \frac{1}{2}\vec{u}.\vec{u}$ in the vortex tube shows a dependence on

- Temporal pressure fluctuations $\frac{\partial p}{\partial t}$
- Heat flux due to the diffusion of temperature $\nabla.(\lambda\nabla T)$
- A source term due to friction $\nabla.(\bar{\tau}.\vec{u})$

$$\rho\frac{d}{dt}h_t = \frac{\partial}{\partial t}p + \nabla.(\lambda\nabla T) + \nabla.(\bar{\tau}.\vec{u}) \tag{4.1}$$

Here $\bar{\tau} = \mu((\nabla u) + (\nabla u)^T - 2/3\mathbb{1}(\nabla.\vec{u}))$ is the stress tensor. It is assumed that the fluid is of Newtonian nature.

This partial differential equation is now examined in detail with its impact on the effect of energy separation in the RHVT. The left-hand side constitutes a simple convection equation. The right-hand side is somewhat more interesting and is therefore examined in detail.

When the balance equation for the total enthalpy 4.1 is integrated over the volume of the vortex tube, the global enthalpy balance 2.3 should be recovered.

First the heat flux

$$\vec{q} = -\lambda\nabla T \tag{4.2}$$

due to temperature difference is investigated by integration on a control volume V.

$$\int_V \nabla.(\lambda\nabla T)\,dV = \int_A (\lambda\nabla T).\vec{n}dA \tag{4.3}$$

The volume integral on the divergence of the heat flux can be replaced by the closed surface integral of the control volume V. Since we assumed an adiabatic condition in Chap. 2 with $\vec{q} = -\lambda\nabla T = 0$, the effect of this term vanishes at the walls.

On the walls, we achieve a similar result concerning the effects due to viscosity. The energy flux due to the viscosity

$$\vec{e}_\nu = -\bar{\tau}.\vec{u} \tag{4.4}$$

receives the same treatment as the heat flux $\vec{q}$. The integral on the control volume of the divergence of energy flux can be expressed as a closed surface integral.

$$\int_V \nabla.(\bar{\tau}.\vec{u})\,dV = \int_A (\bar{\tau}.\vec{u})\,.\vec{n}dA \tag{4.5}$$

Exact application of the integration rules and the boundary condition of $\vec{u} = 0$ at the wall let this term become zero with the exception of the inlet and outlet boundaries. But since inlet and outlet boundary cross sections are small compared with the surface of the wall in the considered geometries, the volume integral of this term can be neglected. As $\nabla.(\bar{\tau}.\vec{u})$ is not zero in the domain, this term must take **positive** as well as **negative** values in the vortex tube to balance the total enthalpy.

As a consequence, there must be regions in the vortex tube with total enthalpy decrease and others with total enthalpy rise.

At this point, the viscous energy flux $-\bar{\tau}.\vec{u}$ requires some more attention. At first glance, it seems to be a simple source term in the total enthalpy equation once the divergence operator has been applied. But, it allows a different interpretation: actually it not only shifts kinetic energy into sensible enthalpy but also does so by generating a flux of energy from one region to a neighbouring region. This finding will be illustrated in a more detailed form when we present the kinetic viewpoint of the physics.

Here it is emphasised that it is a key argument for viscosity being the driving element for the energy separation. This viscosity term was first adopted by Van Deemter [23] in his extended Bernoulli equation to explain the temperature separation.

If the vortex tube enthalpy balance is integrated over a significantly long time scale Δt, we can expect the integral of the pressure fluctuations $\frac{\partial p}{\partial t}$ to vanish as well. Then only the enthalpy fluxes on the inlet and the outlets remain and the global enthalpy balance 2.3 is recovered.

$$\int_V \rho \frac{d}{dt} h_t dV = \dot{H}_i + \dot{H}_c + \dot{H}_h = 0 \tag{4.6}$$

If the acoustic fluctuations due to pressure oscillations induce temperature separation, this would feed the argument of acoustic streaming. Acoustic streaming is not considered further at this point, and so the focus is put on the analysis of the viscous term.

The energy flux due to viscosity $\vec{e}_\nu$ is normally neglected since it does not play a dominant role in most applications. In many CFD codes, it is not even correctly implemented. In the following section, dimensional analysis will be employed to provide justification for these comments.

4.1.2 Dominant Effects in the Total Enthalpy Balance

The dominant effects in the enthalpy balance are most easily isolated by making the balance equation non-dimensional. For all the readers not acquainted with dimensional analysis, just think of it as the universal pliers used by engineers to extract the important physical effects. This is simply done by extracting the effects that are largest in magnitude. Here, the factors of the individual terms give an indication to the importance of the effect. For convenience, the convention of the Mass–Length–Time–Temperature ($MLT\Theta$) system is used. Non-dimensional properties are marked by a star, i.e. non-dimensional length $x^* = x/l_0$ is obtained by division with a reference length and so on for all the other quantities.

$$
\begin{aligned}
x^* &= x/l_0 & & \\
t^* &= t/t_0 & u_0 &= l_0/t_0 \\
h_t^* &= h_t/h_0 & h_0 &= c_p T_0 \\
T^* &= T/T_0 & & \\
p^* &= p/p_0 & p_0 &= \rho_0 R T_0 \\
\rho^* &= \rho/\rho_0 & & \\
\lambda^* &= \lambda/\lambda_0 & & \\
\tau^* &= \tau/\tau_0 & \tau_0 &= \mu_0 u_0/l_0
\end{aligned}
$$

Care must be taken that some dimensional references are linked. The corresponding relations are given in the list above. Introduction of the reference values into the equation of the total enthalpy balance leads to a differential equation with non-dimensional variables.

$$\frac{\rho_0 h_0}{t_0}\rho^*\frac{d}{dt^*}h_t^* = \frac{p_0}{t_0}\frac{\partial}{\partial t^*}p^* + \frac{\lambda_0 T_0}{l_0^2}\nabla^*.\left(\lambda^*\nabla T^*\right) + \frac{\mu_0 u_0^2}{l_0^2}\nabla^*.\left(\bar{\tau}^*.\vec{u}^*\right) \tag{4.7}$$

$$\rho^*\frac{d}{dt^*}h_t^* = \frac{p_0}{\rho_0 h_0}\frac{\partial}{\partial t^*}p^* + \frac{\lambda_0 T_0 t_0}{\rho_0 h_0 l_0^2}\nabla^*.\left(\lambda^*\nabla T^*\right) + \frac{\mu_0 u_0^2 t_0}{\rho_0 h_0 l_0^2}\nabla^*.\left(\bar{\tau}^*.\vec{u}^*\right) \tag{4.8}$$

The non-dimensional coefficients on the right-hand side are analysed individually. The first term preceding the non-dimensional pressure fluctuation $\frac{\partial}{\partial t^*}p^*$ is simply a property of the working fluid and depends on the ratio of the heat capacities $\kappa = c_p/c_v$.

$$\frac{p_0}{\rho_0 h_0} = \frac{\rho_0 R T_0}{\rho_0 c_p T_0} = \frac{R}{c_p} = \frac{\kappa - 1}{\kappa} \tag{4.9}$$

The non-dimensional heat flux due to temperature differences $\nabla^*.\left(\lambda^*\nabla T^*\right)$ is preceded by the Fourier number [26] $Fo = \frac{a_0 t_0}{l_0^2} = \frac{a_0}{l_0 u_0}$. This number gives the time scale, at which diffusion of temperature can smooth out temperature differences.

$$\frac{\lambda_0 T_0 t_0}{\rho_0 h_0 l_0^2} = \frac{\lambda_0 t_0}{\rho_0 c_p l_0^2} = \frac{\lambda_0}{\rho_0 c_p l_0 u_0} = \frac{a_0}{l_0 u_0} = \Pi_1 = Fo \tag{4.10}$$

The non-dimensional friction term $\nabla^* . (\bar{\tau}^* . \vec{u}^*)$ is preceded by the ratio of Eckert to Reynolds number $Ec/Re = \frac{u_0^2}{c_p T_0} \frac{\nu_0}{l_0 u_0}$.

$$\frac{\mu_0 u_0^2 t_0}{\rho_0 h_0 l_0^2} = \frac{\mu_0 u_0}{\rho_0 c_p T_0 l_0} = \frac{\nu_0 u_0}{c_p T_0 l_0} = \Pi_2 = \frac{Ec}{Re} \tag{4.11}$$

The Eckert number is used in forced heat convection comparing the kinetic energy to transferred sensible enthalpy. The Reynolds number usually classifies the type of flow (laminar, turbulent) by comparing inertial to friction forces.

The importance of heat flux compared to friction forces in the total enthalpy balance can be estimated by taking the ratio of the non-dimensional numbers that are preceding the corresponding terms.

$$\Pi_3 = \frac{\Pi_1}{\Pi_2} = \frac{Fo \cdot Re}{Ec} = \frac{\lambda_0}{\rho_0 c_p l_0 u_0} \cdot \frac{\rho_0 c_p T_0 l_0}{\mu_0 u_0} = \frac{\lambda_0 T_0}{\mu_0 u_0^2} \tag{4.12}$$

This non-dimensional number is much larger than unity when the diffusion of temperature is more important than the friction forces $\Pi_3 \gg 1$. In most applications, this is the case. An incorrect or missing implementation of the viscous term in CFD for cases of $\Pi_3 \gg 1$ does not significantly alter the solution of total temperature, since the net contribution of the viscous terms in the total domain is zero.

As examples, typical numbers are added here for water and air at ambient conditions.

Example: Water in a Pipe

$$\lambda_w = 0.6\,\text{W/mK} \qquad \mu_W = 1.0 \cdot 10^{-3}\,\text{kg/ms}$$

$$\Delta T_w = 10\,\text{K} \qquad u_W = 5\,\text{m/s}$$

$$\Pi_{3,W} = \frac{0.6\,\text{W/mK} \cdot 10\,\text{K}}{10^{-3}\,\text{kg/ms} \cdot 25\text{m}^2/\text{s}^2} = 240$$

Example: Air Flow at Ambient Conditions

$$\lambda_A = 0.025\,\text{W/mK} \qquad \mu_A = 2.5 \cdot 10^{-5}\,\text{kg/ms}$$

$$\Delta T_A = 10\,\text{K} \qquad u_A = 5\,\text{m/s}$$

$$\Pi_{3,A} = \frac{0.025\,\text{W/mK} \cdot 10\,\text{K}}{2.5 \cdot 10^{-5}\,\text{kg/ms} \cdot 25\,\text{m}^2/\text{s}^2} = 400$$

In both typical cases, the smoothing effect of temperature diffusion is much larger than the effect of the creation of enthalpy difference due to viscous effects.

Velocity enters the non-dimensional quantity Π_3 as a square in the denominator. This makes the non-dimensional number very sensible to high reference velocities.

4.1.3 *Non-dimensional Value of $Fo \cdot Re/Ec$ for the Vortex Tube*

With typical velocity values for the flow inside the vortex tube, the non-dimensional number Π_3 becomes significantly smaller than one and the viscous effects are dominant compared to the smoothing effect of temperature diffusion as can be seen in the following example:

$$\lambda_{VT} = 0.025\,\text{W/mK} \qquad \mu_{VT} = 2.5 \cdot 10^{-5}\,\text{kg/ms}$$

$$\Delta T_{VT} = 10\,\text{K} \qquad u_{VT} = 300\,\text{m/s}$$

$$\Pi_{3,VT} = \frac{0.025\,\text{W/mK} \cdot 10\,\text{K}}{2.5 \cdot 10^{-5}\,\text{kg/ms} \cdot 90{,}000\,\text{m}^2/\text{s}^2} = 0.11$$

The only difference to the previous example of air at ambient conditions is the assumption of high velocity $u_{VT} = 300$ m/s. This shows that the high swirling velocities are the central ingredient to make the viscous effect become dominant.

Another feature of the non-dimensional number $\Pi_3 = Fo \cdot Re/Ec$ is the independence on length scales, for instance, as pipe diameter. Therefore the effect does occur in pipes of any diameter. To achieve such high velocities in tangential direction in large diameter tubes, large mass volume rates are necessary as the volumetric flow rate scales as the third power of the diameter to achieve a similar flow structure.

It is, however, interesting to note that the temperature separation effect in the vortex tube cannot be reproduced [4] when the viscous term is not correctly implemented.

This requires a more detailed analysis of the viscous terms in the enthalpy equations.

Nomenclature of the Viscous Term

Since the viscous term in the enthalpy equation 4.1 is subject of several further passages, it is given the symbol Φ_{ht}.

$$\Phi_{ht} = \nabla.(\bar{\tau}.\vec{u}) \tag{4.13}$$

This quantity is made non-dimensional by dynamic viscosity μ, velocity u_0 and length scale l_0.

$$\Phi_{ht} = \frac{\mu u_0^2}{l_0^2}\Phi_{ht}^* = \frac{\mu u_0^2}{l_0^2}\nabla^*.\left(\bar{\tau}^*.\vec{u}^*\right) \tag{4.14}$$

The factor $\mu u_0^2/l_0^2$ depends on the square of the characteristic length scale l_0. Small characteristic length scales exhibit a significantly larger viscous source term.

In the next sections, the relations between total enthalpy, sensible enthalpy and kinetic energy with respect to the energy transfer are discussed. This leads us to revisit the equations to define symbols for the corresponding viscous source terms.

4.1.4 Conservation of Kinetic Energy

The equation for kinetic energy can easily be derived from the momentum equation or found in standard text books on fluid dynamics [2, 5, 6].

$$\rho\frac{d}{dt}\frac{1}{2}\vec{u}.\vec{u} = -\nabla p.\vec{u} + (\nabla.\bar{\tau}).\vec{u} \tag{4.15}$$

The last term in the kinetic energy equation can lead to a decrease in kinetic energy due to viscosity and can thus take negative values. Here, this source term is labeled Φ_{kin}.

$$\Phi_{kin} = (\nabla.\bar{\tau}).\vec{u} \tag{4.16}$$

The source term can be made non-dimensional as in the case of the total enthalpy.

$$\Phi_{kin} = \frac{\mu u_0^2}{l_0^2}\Phi_{kin}^* = \frac{\mu u_0^2}{l_0^2}\left(\nabla^*.\bar{\tau}^*\right).\vec{u}^* \tag{4.17}$$

The factor preceding the non-dimensional source term Φ_{kin}^* is identical to the factor preceding Φ_{ht}^*. Still it has to be shown that characteristic length scale l_0 and characteristic velocity u_0 are identical.

4.1.5 Conservation of Sensible Enthalpy

The equation on sensible enthalpy h_s is also to be found in standard text books on fluid dynamics.

$$\rho \frac{d}{dt} h_s = \frac{d}{dt} p + \nabla.(\lambda \nabla T) + \bar{\tau}: (\nabla \vec{u}) \tag{4.18}$$

The last term in the sensible enthalpy equation leads to an increase of sensible enthalpy due to viscosity. This source term is labeled Φ_{hs}.

$$\Phi_{hs} = \bar{\tau}: (\nabla \vec{u}) \tag{4.19}$$

The non-dimensional source term is preceded by the same factor as the previous source terms.

$$\Phi_{hs} = \frac{\mu u_0^2}{l_0^2} \Phi^*_{hs} = \frac{\mu u_0^2}{l_0^2} \bar{\tau}^*: \left(\nabla^* \vec{u}^*\right) \tag{4.20}$$

4.1.6 Links Between the Energy Equations and Viscous Source Terms

As the total enthalpy (h_t) is defined as the sum of sensible enthalpy (h_s) and kinetic energy $(\frac{1}{2}\vec{u}.\vec{u})$

$$h_t = h_s + \frac{1}{2} \vec{u}.\vec{u} \tag{4.21}$$

the same holds true for the corresponding balance equations.

$$\frac{d}{dt} \rho h_t = \frac{d}{dt} \rho h_s + \frac{d}{dt} \rho \frac{1}{2} \vec{u}.\vec{u} \tag{4.22}$$

When the balance equations are added, it is found that the viscous source terms are linked in the same way.

$$\nabla.(\bar{\tau}.\vec{u}) = \bar{\tau}: (\nabla \vec{u}) + (\nabla.\bar{\tau}).\vec{u} \tag{4.23}$$

Using the previously established definitions for the source terms, we get the following simple relation:

$$\Phi_{ht} = \Phi_{hs} + \Phi_{kin} \tag{4.24}$$

As pointed out above, the volume integral on the RHVT of Φ_{ht} must be zero. Whereas Φ_{hs} is always positive, Φ_{kin} may take both positive and negative values. Therefore regions with negative Φ_{ht} can only occur in regions of negative Φ_{kin}.

Moreover, the decrease in kinetic energy Φ_{kin} must dominate the increase in sensible enthalpy by Φ_{hs} when energy separation is to be achieved. This can only take place if a viscous energy flux in the form $-\bar{\tau}.\vec{u}$ transports kinetic energy to a neighbouring region.

Furthermore Eq. 4.24 shows that the source terms share the same characteristic length l_0 and characteristic velocity u_0.

The next interesting step is to identify the regions where the viscous effects in the total enthalpy balance Φ_{ht} have a negative contribution to the enthalpy balance and may lead to a decrease in total enthalpy.

4.2 Dependence of the Viscous Terms on the Velocity Profile

The viscous terms in the total enthalpy and kinetic energy equation are somewhat difficult to imagine. Therefore the effects are exemplified with some simple artificial velocity profiles leading to simple analytical solutions.

In the previous section on the corresponding non-dimensional number, it was shown that laminar flow with small velocities will not exhibit a significant temperature separation, since the effects of temperature diffusion are too large compared to the generating effects of temperature separation. Due to the simplicity of the laminar flow, it is however an instructive start for the investigation on the impact of viscous energy flux $\vec{e}_\nu$ with the Poiseuille flow.[1]

4.2.1 Analytical Example of Viscous Term in a Laminar Pipe Flow (Poiseuille) Profile

Assuming a pipe of radius r_0 with a peak velocity u_0, the laminar flow profile in the axial direction is

$$u_z(r) = u_0\left(1 - \left(\frac{r}{r_0}\right)^2\right) \tag{4.25}$$

[1] Think of it as the streetlight effect: a policeman sees a drunken man searching for something under a streetlight and asks what the drunk has lost. The man answers that he lost his keys, and they both look under the streetlight together. After a few minutes, the policeman asks if he is sure he lost them here, and the drunk man replies, no, that he lost them in the park. The policeman then asks why he is searching here, and the drunk replies 'This is where the light is'.

All other velocity components are neglected. In addition the flow is considered incompressible. The total enthalpy balance for this flow reduces to the following equation:

$$\rho \frac{d}{dt} h_t = \frac{\partial}{\partial t} p + \lambda \frac{1}{r} \frac{\partial}{\partial r} \left(r \frac{\partial}{\partial r} T \right) \tag{4.26}$$
$$+ \mu \left(\frac{u_z}{r} \frac{\partial}{\partial r} \left(r \frac{\partial}{\partial r} u_z \right) + \left(\frac{\partial}{\partial r} u_z \right)^2 \right)$$

As derived before, the viscous term for the total enthalpy balance (Φ_{ht}) may take both positive and negative values.

$$\Phi_{ht} = \mu \left\{ \frac{u_z}{r} \frac{\partial}{\partial r} \left(r \frac{\partial}{\partial r} u_z \right) + \left(\frac{\partial}{\partial r} u_z \right)^2 \right\} \tag{4.27}$$
$$= \mu \frac{4 u_0^2}{r_0^2} \left(2 \left(\frac{r}{r_0} \right)^2 - 1 \right)$$

In the case of the Poiseuille flow, the viscous source term Φ_{ht} takes negative values in the centre of the cylinder ($r = 0$) and positive values at the wall ($r = r_0$).

The viscous term of the kinetic energy balance (Eq. 4.16) for the case of the Poiseuille flow can be computed.

$$\Phi_{kin} = \vec{u}.(\nabla.\bar{\tau}) = \mu \frac{4 u_0^2}{r_0^2} \left(\left(\frac{r}{r_0} \right)^2 - 1 \right)$$

The source term Φ_{kin} takes negative values in the centre of the tube ($r = 0$) and is zero on the wall ($r = r_0$).

The viscous term of the sensible enthalpy balance (Eq. 4.19) is always positive. For the Poiseuille flow, it is just a quadratic function.

$$\Phi_{hs} = \bar{\tau} : \nabla\vec{u} = \mu \frac{4 u_0^2}{r_0^2} \left(\frac{r}{r_0} \right)^2$$

The source term Φ_{hs} is zero on the cylinder axis ($r = 0$) and positive on the wall ($r = r_0$).

The effects of the viscous terms in kinetic energy, sensible enthalpy and total enthalpy for the Poiseuille flow are illustrated in Fig. 4.1. The flow is normalised to $u_{z0} = 1$ m/s, $r = 1$ m and $\nu = 1\ \text{m}^2/\text{s}$. The top graph shows the normalised velocity profile $u^*(r^*)$ of a Poiseuille flow, the second graph shows the source term of kinetic energy dissipation $\Phi^*_{kin}(r^*)$, the third graph shows the dissipative source term in sensible enthalpy $\Phi^*_{hs}(r^*)$ and the bottom graph shows the source term in

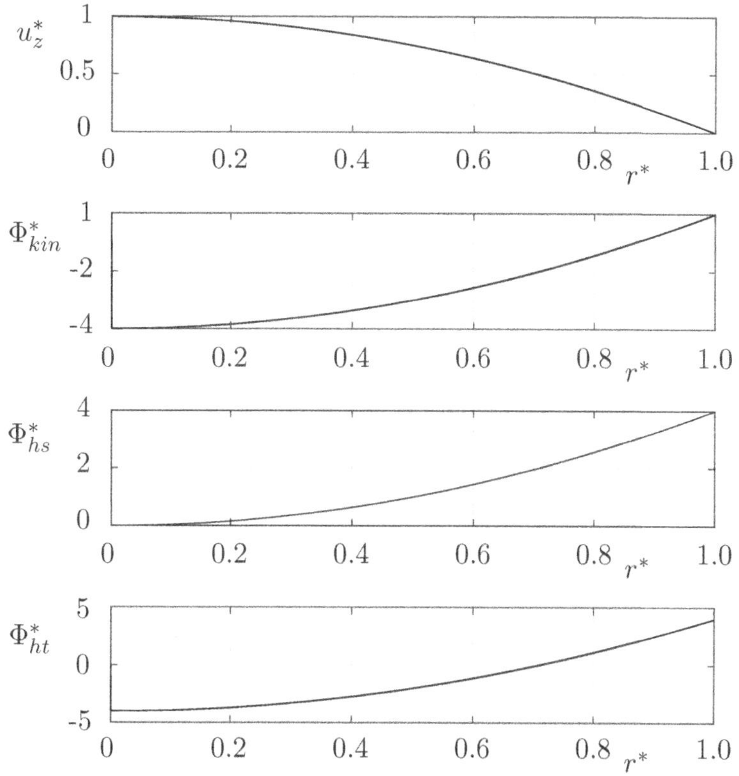

Fig. 4.1 Non-dimensional profiles of laminar velocity u^*, kinetic energy source term Φ_{kin}^*, sensible enthalpy source term Φ_{hs}^* and total enthalpy source term Φ_{ht}^* for the Poiseuille flow

total enthalpy $\Phi_{ht}^*(r^*)$. The tube radius r_0, the velocity u_0 and the viscosity μ_0 were used to make the quantities non-dimensional. The kinetic energy is dissipated across the entire cross section and fed by the pressure gradient in the axial direction to maintain the flow. Due to the viscous term, the total enthalpy decreases on the centreline of the flow and increases towards the wall.

This simple analytical example shows that dissipation of kinetic energy Φ_{kin} is different from the term in the energy balance in the total enthalpy equation Φ_{ht}.

4.2.2 The Viscous Energy Flux in Poiseuille Flow

In the case of the Poiseuille flow, the viscous energy flux in the radial direction can be easily calculated.

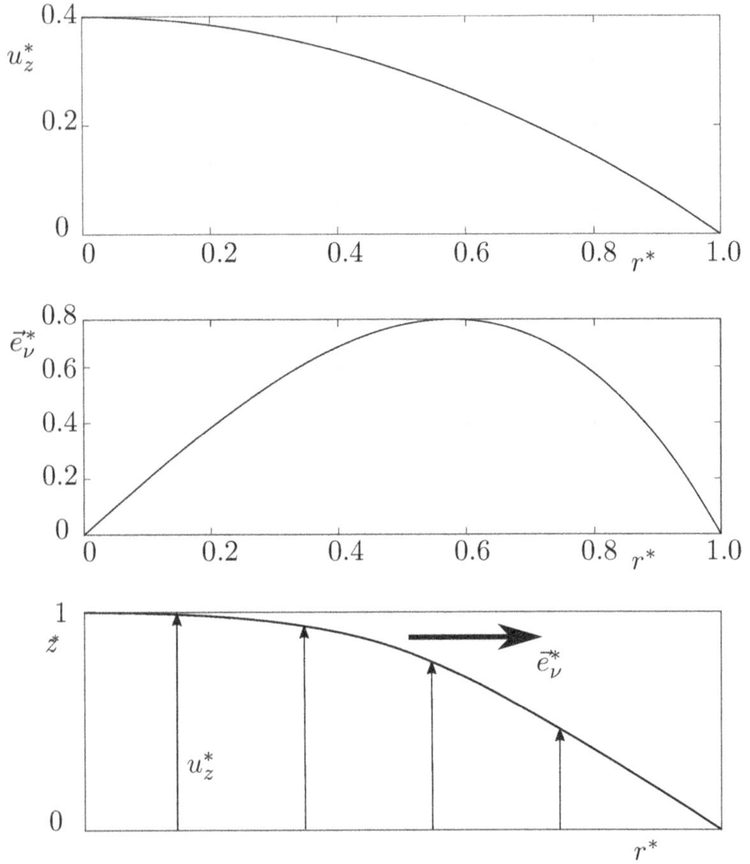

Fig. 4.2 Viscous energy flux in radial direction for the case of Poiseuille flow. Top: non-dimensional velocity profile ($u^*(r^*)$), middle: non-dimensional viscous energy flux in radial direction $\vec{e}_\nu^*$ and bottom: direction of viscous energy flux

$$\vec{e}_\nu = -\bar{\tau}.\vec{u} = -\mu u_z(r)\frac{\partial u_z(r)}{\partial r} \tag{4.28}$$
$$= -\mu 2\left(\frac{u_0^2}{r_0}\right)\left(\left(\frac{r}{r_0}\right) - \left(\frac{r}{r_0}\right)^3\right)$$

This flux shown in Fig 4.2 is positive with a viscous transfer of energy from the centre of the cylinder outwards to the wall. The viscous flux is zero along the symmetry axis and on the cylinder wall. It assumes the maximum value at roughly 60% of the radius and points from the centre of the tube towards the cylinder wall.

The viscous term in the enthalpy equation may therefore be positive in some regions, but negative in others.

4.3 Analytical Example of Viscous Term with a Synthetic Flow Profile

The velocity profiles found in the CFD computations of the turbulent pipe and the RHVT is somewhat different from the Poiseuille flow in the previous example. In particular the strong radial gradients are lacking in the Poiseuille flow example.

Here, a more representative but still synthetic flow profile is analysed with respect to the impact on the viscous term. The synthetic velocity profiles for the axial component u_z and the tangential component u_θ are meant to represent profiles that differ from solid body rotation. They should include some steep gradients next to the walls.

Again the velocity profiles are assumed to depend only on the radial direction r. All other derivatives are assumed to be unimportant. A model equation for both profiles fulfilling the requirements for both velocity components is given by $u_\gamma(r)$.

$$u_\gamma(r) = u_{ref}\left(\frac{r}{r_0}\right)^k\left(1-\left(\frac{r}{r_0}\right)^j\right) \tag{4.29}$$

$$= u_{ref}\left(\left(\frac{r}{r_0}\right)^k - \left(\frac{r}{r_0}\right)^{k+j}\right)$$

$$u_{ref} = u_0 \cdot \left(\frac{j}{k}+1\right)^{k/j}\left(1+\frac{k}{j}\right)$$

The requirements are fulfilled for all non-zero positive integers k and j. Here, for both the axial velocity $u_z(r)$ and the tangential velocity u_θ, the same model equation is applied. The parameters u_0, k and j may of course differ for the velocities u_z and u_θ.

$$u_z(r) = u_{\gamma,1} \qquad u_\theta(r) = u_{\gamma,2} \tag{4.30}$$

It is a simple exercise to compute the contributions of the viscous term in the total enthalpy balance Φ_{ht} due to u_z and u_θ. In this model calculation, we assume $u_r = 0$, $\frac{\partial}{\partial\theta} = 0$ and $\frac{\partial}{\partial z} = 0$. Due to the form of the stress term τ, the contributions of u_z and u_θ are different. Therefore the contributions are computed individually.

$$\Phi_{ht,u_z} = \mu\left(\frac{u_\gamma}{r}\frac{\partial}{\partial r}\left(r\frac{\partial}{\partial r}u_\gamma\right) + \left(\frac{\partial}{\partial r}u_\gamma\right)^2\right)$$

$$= \mu\frac{u_{ref}^2}{r_0^2}\left(2k^2\left(\frac{r}{r_0}\right)^{2k-2} + 2(j+k)^2\left(\frac{r}{r_0}\right)^{2(j+k)-2}\right.$$

$$\left. -(2k+j)^2\left(\frac{r}{r_0}\right)^{2k+j-2}\right)$$

$$\begin{aligned}\Phi_{ht,u_\theta} &= \frac{1}{r}\frac{\partial}{\partial r}\left(r\left(\frac{\partial u_\theta}{\partial r} - \frac{u_\theta}{r}\right)u_\theta\right)\\ &= \mu\frac{u_{ref}^2}{r_0^2}\left((2k^2-2k)\left(\frac{r}{r_0}\right)^{2k-2}\right.\\ &\quad + (2(j+k)^2 - 2(k+j))\left(\frac{r}{r_0}\right)^{2(j+k)-2}\\ &\quad \left. -((2k+j)^2 - 2(2k+j))\left(\frac{r}{r_0}\right)^{2k+j-2}\right)\end{aligned}$$

Taking the same choices j and k for the velocities u_z and u_θ, the contributions of the source term differ for u_z and u_θ.

The source term for the synthetic profiles is given in Fig 4.3. Two effects can be noticed. First, this velocity profile shows negative values for the total enthalpy source term outside the boundary layer of the wall. Second, the non-dimensional values for the source term increase with the magnitude of the gradients. Tubes with small diameter and large circumferential velocities u_θ show significantly larger source terms due to larger gradient values.

The viscous energy flux $\vec{e}_\nu$ is computed separately for the two contributions: the contribution due to u_z: $\vec{e}_{\nu,u_z}$ and the contribution due to u_θ: $\vec{e}_{\nu,u_z}$.

$$\vec{e}_\nu = -\bar{\tau}.\vec{u} = \underbrace{-\mu u_\theta\left(\frac{\partial u_\theta}{\partial r} - \frac{u_\theta}{r}\right)}_{\vec{e}_{\nu,u_\theta}} \underbrace{-\mu u_z\left(\frac{\partial u_z}{\partial r}\right)}_{\vec{e}_{\nu,u_z}} \tag{4.31}$$

The contributions for the radial viscous energy fluxes for identical choices of $k = 2$ and $j = 5$ are shown in Fig. 4.4.

Both flux contributions show transfer regions of viscous energy. In a range of non-dimensional radius of approximately 0.4–0.8, the viscous energy flux is negative and transporting energy from the region to the centre of the tube. In the outer part with non-dimensional radii larger than 0.8, the viscous energy flux is positive and transports energy from this region to the tube wall.

4.4 Effect of the Viscous Term in the Turbulent Pipe with Rotating Flow

In Chap. 3, velocity profiles of the turbulent pipe with swirling flow were shown. Here we show the corresponding non-dimensional source terms for

- Kinetic energy Φ_{kin}
- Sensible enthalpy Φ_{hs}
- Total enthalpy Φ_{ht}

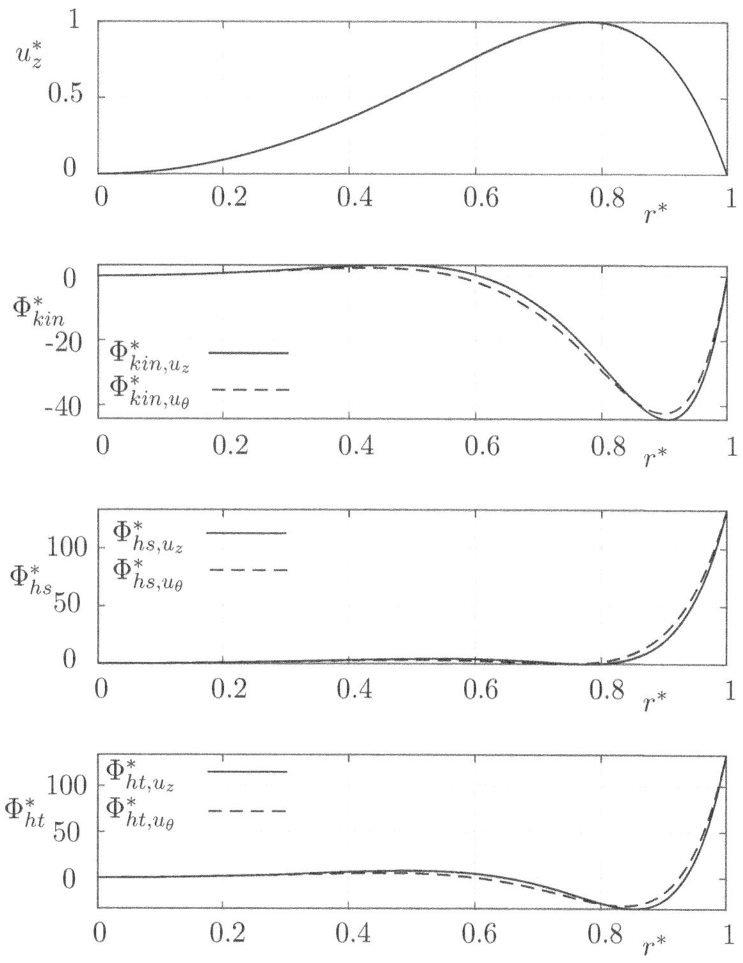

Fig. 4.3 Non-dimensional synthetic profiles of velocity with $k = 2$ and $j = 5$, kinetic energy source term Φ_{kin}^*, sensible enthalpy source term Φ_{hs}^* and total enthalpy source term Φ_{ht}^*

Source terms are normalised by pipe radius r_0, velocity u_0 and viscosity μ_0. All the source terms play an important role at the inlet boundary condition and most of all at the cylinder wall. Directly on the cylinder wall, the viscous source term Φ_{ht}^* is positive and of large magnitude. On the layer just inside the boundary layer, the source term is not only negative but also of large magnitude.

For a comparison to the source terms obtained from the synthetic velocity profile, the source terms are evaluated from the mean flow field and shown as a function of radius in Fig. 4.5. The viscous source terms are most of all active in the boundary layer. With respect to the source terms, this layer can be divided into an outer layer and an inner layer. In the outer layer, the source term in the sensible energy is of

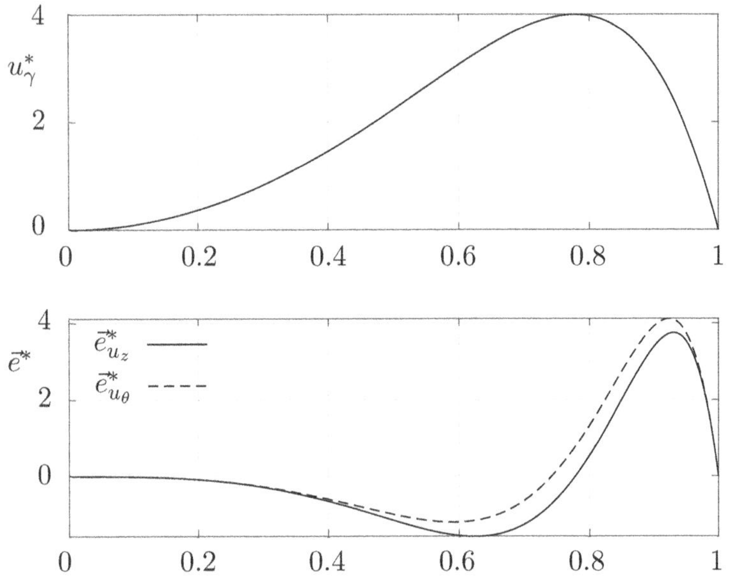

Fig. 4.4 Non-dimensional radial energy fluxes for the synthetic velocity profiles

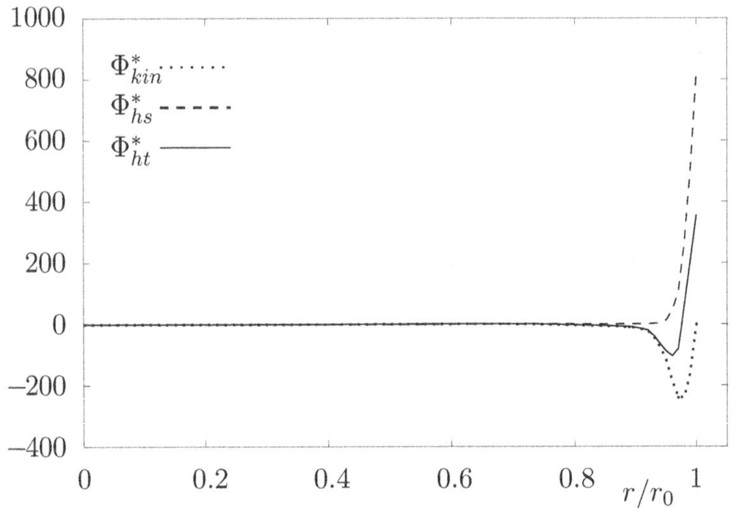

Fig. 4.5 Non-dimensional source terms obtained from mean flow field at $z = 0.5l/d$, kinetic energy source term Φ^*_{kin}, sensible enthalpy source term Φ^*_{hs} and total enthalpy source term Φ^*_{ht}

largest magnitude. In the inner layer, the kinetic energy source term is dominant. The source term of kinetic energy balance and sensible enthalpy h_s sum up to the source term of total enthalpy h_t.

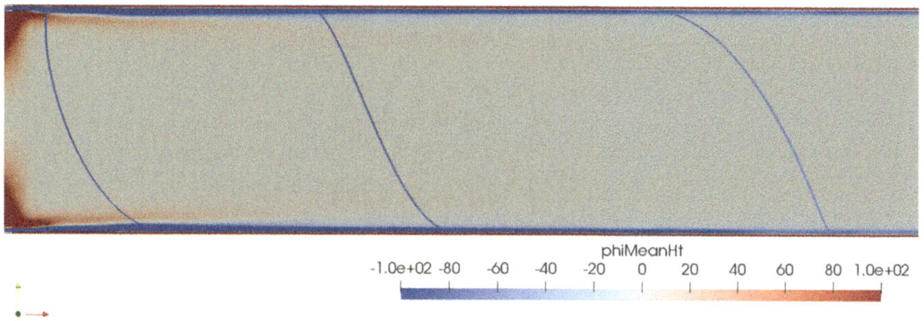

Fig. 4.6 Streamline obtained from the time-averaged flow field travelling through the inner boundary layer

Thought Experiment on Streamlines

If a fluid particle travels most of its way through the region with a negative source term in h_t, here the inner boundary layer, the total enthalpy of the fluid particle will decrease. On the other hand, a fluid particle travelling in the outer boundary layer with a positive source term in h_t will see an increase in total enthalpy.

This is nothing but a handwaving argument. It would be necessary to track fluid particles from the inlet to the outlet and to integrate over the source terms along the fluid particle path to get a quantitative answer.

Streamlines are a dangerous tool to use to interpret flow as they can give misleading indications. Nevertheless, in the next step, streamlines obtained from the mean flow fields are used to examine whether or not fluid particles travel long enough through zones of negative Φ_{ht} to obtain a significant decrease of total enthalpy.

In Fig. 4.6, such a streamline obtained from a mean field is shown. Since the streamline is obtained from a mean field, it would be wrong to postulate that the same holds true to a fluid particle in a turbulent boundary. It is still interesting to obtain the average source term of a particle on such a streamline. Actually the integration along the streamline is an integration over a length.

$$\frac{1}{l}\int_0^l f dx$$

However we are interested in the residence time of the fluid particle at the particular location. Therefore the length integration is replaced by a time integration.

$$\frac{1}{\tau}\int_0^\tau f dt$$

For the case of the swirling flow, the non-dimensional mean source terms were integrated along the streamline.

$$\frac{1}{\tau}\int_0^\tau \bar{\Phi}^*_{Kin}dt = -462$$
$$\frac{1}{\tau}\int_0^\tau \bar{\Phi}^*_{hs}dt = 259$$
$$\frac{1}{\tau}\int_0^\tau \bar{\Phi}^*_{ht}dt = -183$$

As expected, there is a strong decrease in kinetic energy, an increase in sensible enthalpy and a decrease in total enthalpy. The values can be made dimensional by multiplication with $\mu\frac{u_0^2}{d_0^2}$. A multiplication with the residence time in the tube Δt leads to the decrease in total enthalpy and therefore in total temperature.

Quantitative Evaluation of Thought Experiment
A quantitative estimation of the viscous effect can be carried out by using some typical values. Here the dynamic viscosity $\mu = \rho\nu$, the circumferential velocity u_0, tube diameter d_0, non-dimensional average source term $\bar{\Phi}^*_{ht}$, residence time Δt and the heat capacity of the fluid c_p need to be specified.

$$\mu = 2.0 \cdot 10^{-5}\,\text{kg/ms}$$
$$u_0 = 200\,\text{m/s}$$
$$d_0 = 0.01\,\text{m}$$
$$\bar{\Phi}^*_{ht} = -100$$
$$\Delta t = 0.02\,\text{s}$$
$$c_p = 1005\,\text{J/kgK}$$

For a calorically perfect gas $h_t = c_p T_t$, this leads to a total temperature drop of roughly 16 K in air for a fluid particle travelling through the zone of a negative source term of total enthalpy.

Furthermore this gives, as a first estimate, a linear relation between pressure difference and temperature difference.

$$\Delta T \propto \nu \frac{u_\theta^2}{c_p d_0^2}\bar{\Phi}^*_{ht}\Delta t \tag{4.32}$$

When the equation is divided by a reference temperature, essentially the result of van Deemter [23] with a Mach number $Ma = \sqrt{\frac{u_\theta^2}{c_p T_0}}$, based on the circumferential velocity, is recovered.

$$\frac{\Delta T}{T_0} \propto \nu \frac{Ma_{u_\theta}^2}{d_0^2}\bar{\Phi}^*_{ht}\Delta t \tag{4.33}$$

Up to now only arguments for the turbulent pipe have been given. In the next section, the flow profiles from the RHVT of the chapter on flow in the vortex tube are used to analyse the source terms.

4.5 Effect of the Viscous Term in the Vortex Tube

Next the effect of the viscous term will be discussed as applied to the geometry of Farouk et al. [10, 11] since it can be considered a representative example for most counter-flow RHVT. The flow structure of the geometry was illustrated in Sect. 3.2.4. Here it is shown that the strong velocity gradients in the boundary layer lead to strong source terms for the enthalpies in the vortex tube.

4.5.1 Viscous Source Terms in the RHVT

The viscous source terms are dominant in the regions of high velocity gradients. Since the source terms for the axial velocity component are of one magnitude lower than the tangential velocity component, the following illustrations are limited to the contributions of u_θ. The source terms for the enthalpy equations were computed by postprocessing the time-averaged u_θ velocity field.

$$\begin{aligned}
\Phi_{kin,u_\theta} &= \mu u_\theta \left[\frac{2}{r} \left(\frac{\partial u_\theta}{\partial r} - \frac{u_\theta}{r} \right) + \frac{\partial}{\partial r} \left(\frac{\partial u_\theta}{\partial r} - \frac{u_\theta}{r} \right) \right] \\
\Phi_{hs,u_\theta} &= \mu \left(\frac{\partial u_\theta}{\partial r} - \frac{u_\theta}{r} \right)^2 \\
\Phi_{ht,u_\theta} &= \mu \frac{1}{r} \frac{\partial}{\partial r} \left(r \left(\frac{\partial u_\theta}{\partial r} - \frac{u_\theta}{r} \right) u_\theta \right)
\end{aligned} \tag{4.34}$$

Those source terms are illustrated in Fig. 4.7. The figures show a zoom on the boundary layer. Outside the boundary layer, the source terms are of smaller value.

4.5.2 Viscous Enthalpy Flux

One of the central questions is whether the total enthalpy flux is sufficient to shift the enthalpy from one part of the flow to the other. The total enthalpy flux due to the tangential velocity was computed from the time-averaged u_θ velocity field by

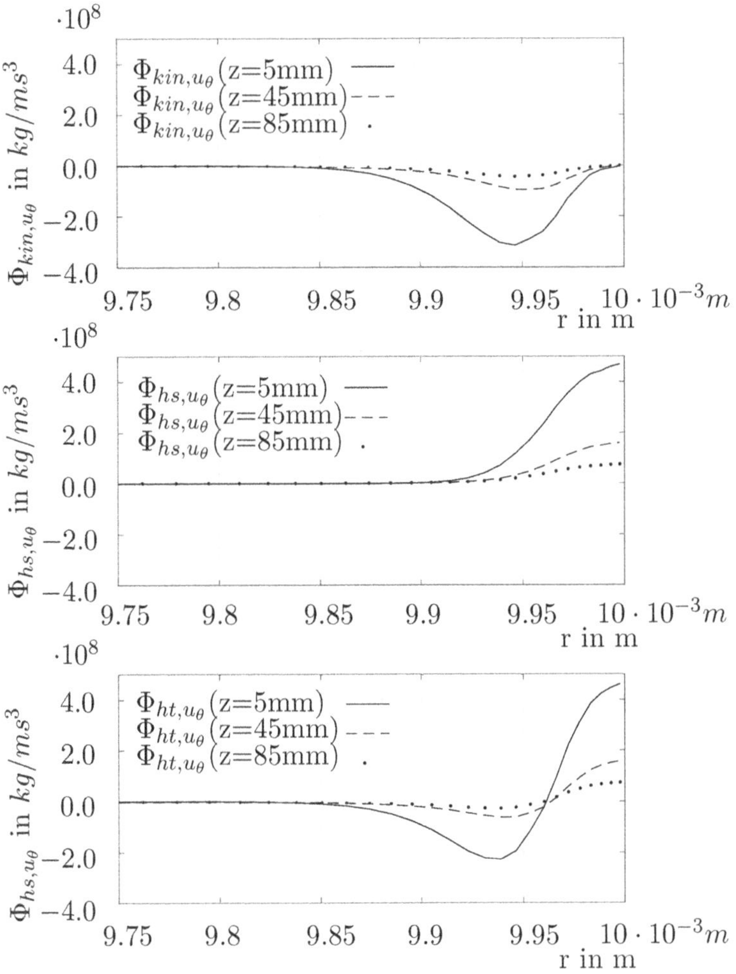

Fig. 4.7 Viscous source terms Φ_{Kin} (top), Φ_{hs} (middle) and Φ_{ht} (bottom) computed from the average u_θ velocity profile in the RHVT

$$\vec{e}_v = -\mu u_\theta \left(\frac{\partial u_\theta}{\partial r} - \frac{u_\theta}{r} \right) \tag{4.35}$$

The result is illustrated in Fig. 4.8. In the next section on thermodynamic estimates, the quantitative values will be used to check if it is plausible that the enthalpy flux is sufficient to transfer the energy.

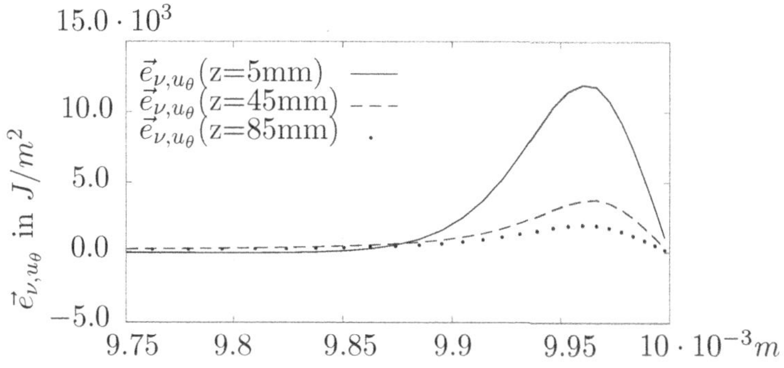

Fig. 4.8 Viscous enthalpy flux in radial direction due to u_θ $\vec{e}_{\nu,\theta}$, computed from the time-averaged u_θ velocity profile in the RHVT

4.5.3 Thermodynamic Estimates

The thermodynamic properties of the boundary conditions with the Farouk tube can be estimated. The intention is to compare the values to the source terms and energy fluxes obtained from the CFD calculations.

The volumetric flow rate at the inlet is easily computed by the radial inlet velocity and the cross section $\dot{V}_i = 0.0019\,\mathrm{m}^3/\mathrm{s}$. Since the CFD calculation is done with constant density, the outlet flow rate is simply computed with the mass flow fraction $\mu = 0.3$ to $\dot{V}_c = \mu \dot{V}_i = 0.00063\,\mathrm{m}^3/\mathrm{s}$.

In the next step, the global enthalpy balance is used to estimate the heat flux needed to separate the hot and the cold flow. The total enthalpy is assumed to correspond to a total temperature of 300 K. The high inlet velocities reduce the sensible enthalpy to a temperature corresponding to 255 K. The energy flux necessary to reduce cold outlet total temperature by 30 K can then be computed by the difference in enthalpy flux $\Delta \dot{H} = \dot{m}_c c_p \Delta T \approx 19\,\mathrm{W}$. This value is only an order of magnitude since the density is not included correctly.

It is interesting to compare this result to the viscous total enthalpy flux near the wall tube. The total flux is obtained by integrating the viscous enthalpy flux on a cylinder surface of the corresponding diameter near the wall. With an average value estimate of $e_{\nu,r,\theta} = 4000\,\mathrm{W/m}^2$ from Fig. 4.8. This yields an estimate for the total viscous energy flux.

$$\dot{E} = \int \vec{e}_{\nu,\theta} 2 \cdot \pi r dl \approx 25\,\mathrm{W} \tag{4.36}$$

This is the same order as the estimate for the necessary difference in enthalpy flux for temperature reduction of roughly 30 K at the cold orifice.

4.6 Transport Equations for Sensible Enthalpy

To illustrate the effect of the source term on the enthalpy balance, a transport equation for sensible enthalpy h_s is added into the *pisoFoam* solver. In the LES computation, the subgrid effects are taken into account by making use of effective thermal conductivity λ_{eff} and effective viscosity μ_{eff}.

$$\rho \frac{d}{dt} h_s = (\nabla p).\vec{u} + \nabla.\left(\lambda_{eff} \nabla T\right) + \bar{\tau} : \left(\nabla \vec{u}\right) \tag{4.37}$$

Terms on temporal pressure oscillation $\frac{\partial p}{\partial t}$ are omitted from the equation. In addition to transport and volumetric compression $(\nabla p.\vec{u})$, enthalpy can only change due to diffusion $(\nabla.\left(\lambda_{eff} \nabla T\right))$, which should create a smoothing characteristic, and viscous effects that generate an increase in sensible enthalpy. The simultaneous solution of the passive transport equation should then indicate the areas of energy separation and deliver different enthalpies at the hot and the cold exit. The total specific enthalpy $h_t = h_s + \frac{1}{2}\vec{u}.\vec{u}$ is then computed from the sensible enthalpy h_s field and the velocity field $\vec{u}$.

4.6.1 *Solution for the Enthalpy Equation*

A cross section of the time-averaged sensible enthalpy, h_s, is shown in Fig. 4.9. The sensible enthalpy reaches significantly higher values on the boundary layer of the wall and diminishes in the centre of the tube.

When the kinetic energy $\frac{1}{2}\vec{u}.\vec{u}$ is added to sensible enthalpy h_s, the total enthalpy is recovered. A cross section of the time-averaged total enthalpy is shown in Fig. 4.9. The total enthalpy is smaller in centre of the tube, leading to lower temperatures at the exit of the cold exit of the RHVT. Instantaneous enthalpy profiles at the different sections are shown in Fig. 4.10.

4.6.2 *Choice of the Characteristic Length Scale*

So far the tube radius r_0 and diameter d_0 were taken as reference values to make the quantities non-dimensional. The preceding examples, however, lead us to the question whether this is the correct choice or not since the source terms occur dominantly in the boundary layer of the flow. The factor making the source terms (see Eq. 4.14) non-dimensional is

$$\frac{\mu u_0^2}{l_0^2} \tag{4.38}$$

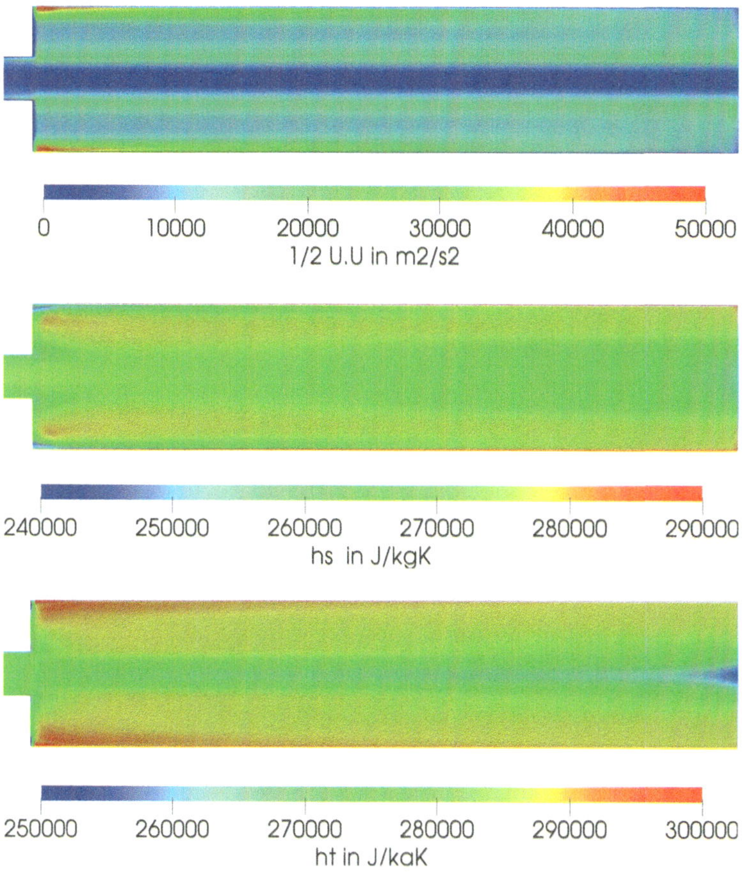

Fig. 4.9 Top: time-averaged kinetic energy $\frac{1}{2}\vec{u}.\vec{u}$ Middle: time-averaged sensible enthalpy h_s Bottom: time-averaged total enthalpy h_t in the RHVT

Here the length scale l_0 arises from the derivatives. As seen in the diagrams on the source terms (Figs. 4.1, 4.3 and 4.5), steep derivatives and strong source terms occur in the boundary layer of the flow. Therefore the thickness of the boundary layer might be a good choice for the corresponding length scale.

The thickness of the turbulent boundary layer was already discussed in the chapter on the flow in the vortex tube, see Eq. 3.11. If this boundary layer thickness is introduced into the scale factor for the source terms, it yields

$$\rho \nu^{4/5} u_0^{11/5} \frac{1}{0.37 \cdot x^{4/5}} \tag{4.39}$$

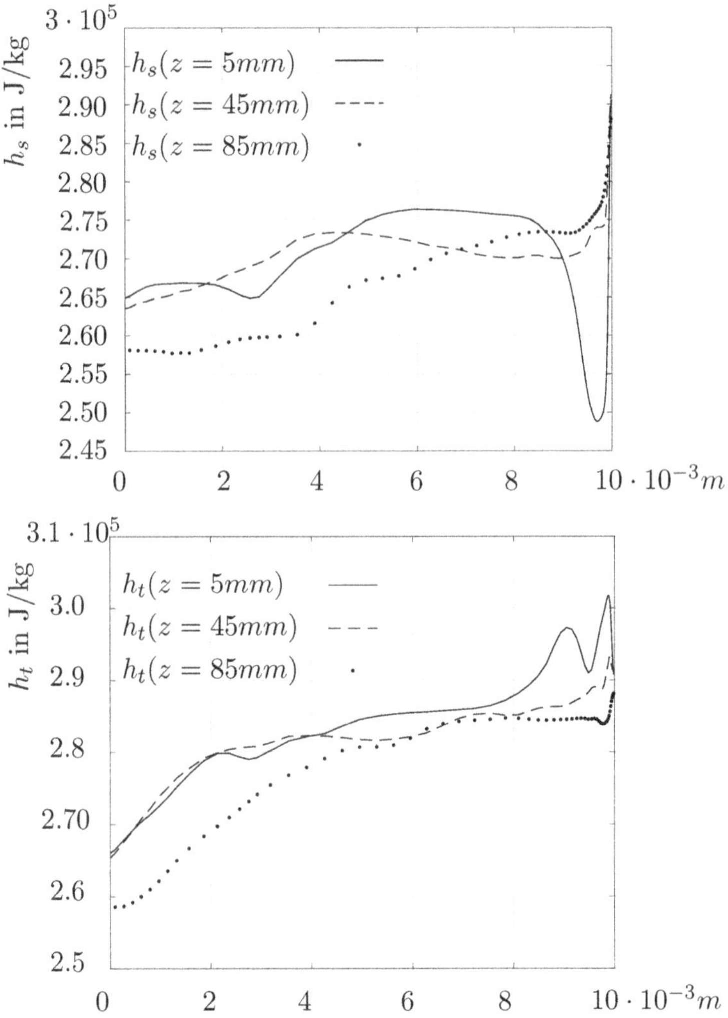

Fig. 4.10 Sensible enthalpy h_s and total enthalpy h_t along the predefined sections, see Fig. 3.8

Here x is the length of the flow path subsequent to the entry of the tube. The dependence of the scaling factor on the flow properties can then be used to establish correlations to estimate the performance of the RHVT.

4.7 Exchange of Kinetic Energy and Sensible Enthalpy

The term $\nabla p.\vec{u}$ appears with opposite sign in the kinetic energy balance (Eq. 4.15) and the sensible enthalpy equation (Eq. 4.18) where it is hidden in the temporal change of the pressure $\frac{d}{dt}p$. This term is known for shifting sensible enthalpy to kinetic energy and back. An illustrative example is the streamline in a nozzle: when the area contracts $\frac{\partial A}{\partial s} < 0$, continuity forces the velocity to increase along the streamline $\frac{\partial u}{\partial s} > 0$. Conservation of total energy requires the pressure to decrease along the streamline $\frac{\partial p}{\partial s} < 0$. Then $\nabla p.\vec{u} < 0$ since the pressure gradient points in the opposite direction of the velocity. This makes the sensible enthalpy h_s decrease and the kinetic energy $\frac{1}{2}\vec{u}.\vec{u}$ increase.

In the case of the vortex tube, the radial pressure gradient is positive $\frac{\partial p}{\partial r} > 0$. To reach the cold outlet in the centre of the tube, there must be a contribution of $u_r < 0$, which leads to $\nabla p.\vec{u} < 0$. Therefore it is legitimate to expect an increase in kinetic energy and a decrease in sensible enthalpy towards the centre of the tube. A comparison in magnitude shows that the viscous contribution in the boundary layer is much larger than this effect. On the other hand, the volume in which $\nabla p.\vec{u}$ has a contribution is much larger than the boundary layer in which the viscous effects occur. So both effects play an important role in the energy balance.

When the kinetic energy is shifted towards sensible enthalpy by $\nabla p.\vec{u}$ and vice versa, this has no impact on the total enthalpy whatsoever. Only the viscous flux in the total enthalpy h_t is shifting the energy from the inner layer to the wall layer.

4.7.1 *Circumferential Kinetic Energy Balance*

It was previously stated that the total enthalpy has to be conserved. This begs the question why the total enthalpy decreases to the centre of the tube as shown in Fig. 4.10. The cause can be found in the kinetic energy. Here, the circumferential velocity in particular is the cause. The circumferential velocity must become zero for an axisymmetric velocity profile as seen in the time-averaged velocity profile in Fig. 3.14.

In the centre of the tube, the time-averaged circumferential velocity profile is almost a solid body rotation. In such a flow, the viscous part vanishes from the momentum equation. As discussed in Sect. 3.3 for the balance of the circumferential velocity, the remaining terms from the averaged equations do not allow the balancing of the equations with the exception of the fluctuating contributions. This is confirmed by the LES computation showing large changes in all velocity contributions in the centre of the tube. This is especially true for the section next to the hot outlet as illustrated in Fig. 3.12.

Considering only the dominant contributions next to the hot outlet and neglecting derivatives in circumferential direction, the kinetic energy equation reduces to the following balance equation in cylinder coordinates:

$$\frac{\partial}{\partial t}\frac{1}{2}\left(u_r^2+u_\theta^2\right)+u_r\frac{\partial}{\partial r}\frac{1}{2}\left(u_r^2+u_\theta^2\right)=-\frac{u_r}{\rho}\frac{\partial}{\partial r}p \tag{4.40}$$

A steady state balance equation would require the second term from the left-hand side to balance the right-hand side of Eq. 4.40. As previously discussed, this is not possible and only unsteady solutions are possible. The fluctuating kinetic energy can be identified in the transient simulations of the vortex tube. Transient visualisations of kinetic energy can be found in in the supplementary material. The LES velocity profiles show a region of higher circumferential velocity at a radius of the order of the outlet orifice. It may be assumed that the fluctuating contributions lead to a shift of circumferential momentum to this region and are the cause of pressure oscillations.

4.8 The Entropy Balance

Another frequently asked question is the compatibility with entropy production: Mischner et al. [14] have published an article on this topic. Their central question is: how can a decrease in total enthalpy be consistent with an increase in entropy without exchange of work or heat to the outside?

The entropy equation shows two mechanisms on changing entropy.

$$\rho T\frac{d}{dt}s=\nabla.(\lambda\nabla T)+\bar{\tau}:(\nabla u) \tag{4.41}$$

The first mechanism is heat transfer by diffusion of temperature. It increases entropy by heat transfer from hot to cold. In the case of the RHVT, this is an increase in entropy on the cold side.

The second mechanism is entropy production by viscous forces $\dot{S}_j$ is $\bar{\tau}:(\nabla u)$. This source term is identical with the source term in the equation for sensible enthalpy h_s (Eq. 4.18) and is therefore always positive. In the case of the RHVT, this is an increase of entropy especially in the regions of hot flow. The entropy balance, Eq. 4.41, does not show any other mechanisms of entropy transfer but the two discussed above, but as long as the entropy of the fluid leaving the hot outlet has taken the largest part of the entropy increase, this is fully compatible with thermodynamics.

In Fig. 4.11, a cross section of the RHVT colour-coded with entropy is shown. Entropy was computed by the transport Eq. 4.41 with a fixed value of 0 J/kgK at the inlet. The entropy increase along the wall can clearly be identified and is within the ranges of a thermodynamic rule of thumb calculation. In a thin layer on the wall, not quite visible in Fig. 4.11, entropy is significantly larger.

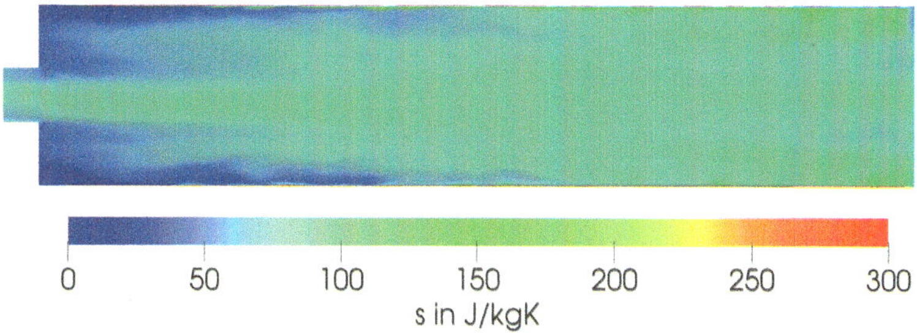

Fig. 4.11 Specific entropy s in the RHVT

4.9 Energy Transfer in Liquid Fluids

Some authors claim to have seen temperature separation in liquid fluids as well [3]. Since the enthalpy equations only change by the equations of state, the previous analysis on ideal, caloric, perfect gases can be extended to liquids. In principle the temperature separation can take place in a liquid as well. The non-dimensional numbers developed before can quickly point to as to what liquids could be good candidates.

The source terms in the enthalpy equations scale by $\mu u^2/l^2$. Since the dynamic viscosity contains the factor of density, the magnitude of the dynamic viscosity of water is three orders of magnitudes larger than the dynamic viscosity of air. On the other hand, the thickness of the boundary layer δ and $y+$ scale by the kinematic viscosity ν. For the same magnitude of velocity gradients, the velocity magnitude must remain similar. However, the typically large densities of liquid fluids require tremendous pressure to accelerate to the velocities necessary to achieve the gradients that are the cause of the energy transfer mechanism. To achieve the velocity of the example from air $u = 300\,\text{m/s}$ in the case of water, an estimate with the Bernoulli equation requires a pressure drop of $\Delta p = 450\,\text{bar}$.

4.10 Energy Transfer Summary

Starting from the balance equation of specific total enthalpy and certain velocity profiles, the viscous radial energy flux in circulating flow could be identified as the transport mechanism shifting kinetic energy from areas of large u_θ to the wall layer where it is then dissipated and turned into sensible enthalpy. The crucial point is that this is actually an energy flux from the area of high circumferential velocity u_θ into the area of high friction at the tube wall. This transfer originates from the strong velocity derivatives in the boundary layer of the tube.

Chapter 5
The Working Mechanism of the RHVT

> The emotion at the point of technical breakthrough is better than wine, women and song put together.
>
> Richard Hamming

The previous chapters focused on thermodynamics, fluid dynamics and energy transport. In this chapter the details are finally put together to get the general picture of the working mechanism of the RHVT. Some of the scientific exactness is omitted to permit a more qualitative understanding of the mechanism.

5.1 Definition of Thermodynamic States

In this section the different thermodynamic states of the flow in the RHVT are (re)defined.

- Reservoir state (1):
  The gas is in a state of high pressure p_r at ambient temperature T_1. Flow velocities are very small $\vec{u}_1 \approx 0$. The total enthalpy is known $h_{t,1} = c_p T_1$ and equals sensible enthalpy $h_{s,1} = h_{t,1}$.
- Nozzle state (2):
  The gas has accelerated to the high velocity of the nozzle $\vec{u}_2$. To achieve this acceleration, pressure decreased significantly and, as a consequence, there is a significant decrease in temperature T_2. Assuming adiabatic flow, the total enthalpy remains constant $h_{t,2} = h_{t,1}$. Sensible enthalpy has decreased due to the high velocity $h_{s,2} < h_{s,1}$.

A, Kaufmann, *The Ranque Hilsch Vortex Tube Demystified*,
https://doi.org/10.1007/978-3-030-89766-6_5

- Tube state (3):
  The flow from the nozzle was directed into the tube to achieve a high circumferential velocity $u_{\theta,3}$. A complex flow pattern develops inside the vortex tube with high tangential and axial velocities next to the tube wall. The high velocity gradients next to the outer wall are really remarkable here. Due to the high velocity, the sensible enthalpy is still significantly smaller than the total enthalpy. $h_{s,3} < h_{t,3}$.
- Exit state hot (4h):
  The viscous effects have increased total enthalpy in the flow next to the wall. The flow next to the tube wall is leaving the tube with the remaining circumferential velocity $u_{\theta,4h}$ and a moderate radial velocity $u_{r,4h}$ at the outside boundary of the tube. The viscous energy transport mechanism has turned kinetic energy from the part of the boundary layer more distant to the wall into sensible enthalpy in the near wall boundary layer. Thus enthalpies are larger than at the intake state: $h_{t,4h} > h_{t,3}$ and $h_{s,4h} > h_{s,3}$.
- Exit state cold (4c):
  The viscous energy transport mechanism has decreased total enthalpy in the flow more distant to the wall $h_{t,4c} < h_{t,3}$. Thus temperatures at the cold exit are smaller than at the hot exit. This flow is leaving the tube with moderate circumferential velocity $u_{\theta,4c}$ and a significant axial velocity $u_{z,4c}$.

These states are illustrated in an h-s diagram in Fig. 5.1. Up to state 3 this is standard thermodynamics. The interesting part is happening between states 3 and 4 where the energy is transferred.

5.2 Molecular Point of View of Momentum Transfer

To give a different, rather handwaving viewpoint on the transfer of energy, a sketch in Fig. 5.2 illustrates the molecules in the near wall region.

The layer of molecules close to the wall is named H and the layer more distant to the wall is named C. Next to the wall (H) the correlated velocity is lower than in the layer more distant from the wall (C). Initially both may have a similar, uncorrelated velocity and therefore a similar temperature. Fast molecules from layer (C) keep bumping into molecules of layer (H) and transfer thereby parts of the kinetic energy. The dominant transfer is from layer (C) to layer (H), since the absolute total velocity of molecules in layer (C) is higher. In both layers, the molecules keep on colliding and ensure an uncorrelated velocity distribution of the Maxwell type. The kinetic energy transfer from layer (C) to layer (H) decreases the total enthalpy in layer (C) and increases total enthalpy in layer (H).

With increasing flow length a significant part of the kinetic energy from the circumferential velocity u_θ in layer C is transferred to uncorrelated velocity, which is in fact temperature, in layer H. The flow in layer H leaves the tube at the hot end.

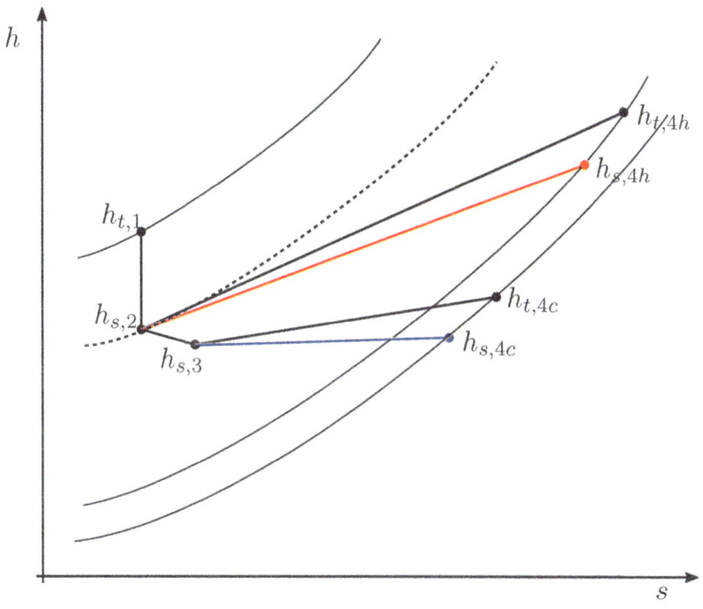

Fig. 5.1 The different thermodynamic states in the RHVT

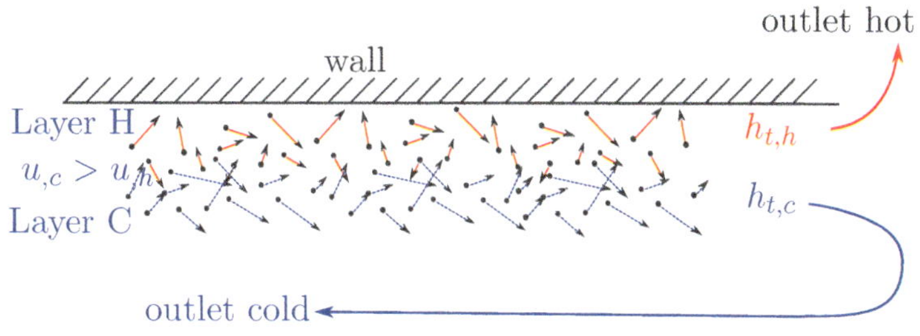

Fig. 5.2 Sketch of momentum transfer in the RHVT

The flow structure forces the flow in layer C to reverse the axial direction to leave at the cold end of the tube.

This corresponds to the flow structure shown in Fig. 3.15. If boundary conditions are set for the case of small cold mass flow ratio μ, only a small percentage of the flow passes by the back wall to the center of the tube. By viscous energy transfer, a significant part of the kinetic energy was passed to the neighbouring hot flow that already left the tube. This part then spirals back to the cold exit. When the cold mass fraction μ becomes dominant, parts of mass flow that received the kinetic energy from the inner layer is also directed by to the center of the tube and leave the tube on the cold outlet. This causes the temperature of the cold flow fraction to rise.

5.3 Working Mechanism Summary

Putting all information together, the temperature decrease at the cold end results from the isentropic expansion in the nozzle and the transfer of kinetic energy to the part of the flow leaving the flow at the hot end. This is largely done just by the viscous kinetic energy flux resulting from the steep velocity gradients. Another important mechanism is transport of kinetic energy by strong velocity fluctuations. Those fluctuations are necessary for the flow to fullfill the circumferential velocity balance in the RHVT.

Chapter 6
Useful Applications of the RHVT

The poor coefficient of performance found in the thermodynamic chapter and the necessity for significant pressure ratios make the RHVT not very attractive for a standard cooling device. For such applications, coolers based on standard vapour compression cycles are far more competitive.

Paul Dirac proposed the vortex tube for isotope separation. This proposal leads to the *Helikon vortex separation process* used by *The Uranium Enrichment Corporation of South Africa* to separate Uranium isotopes. However, the low level of efficiency led to the project being discontinued.

Some authors report on the separation of gases using vortex tubes. But here too, the low level of efficiency means that this technology is not used in practise.

Still, there are special applications, less susceptible to the coefficient of performance in which other features are more important. In the case of the cooling of milling tools, the application of a vapour compression cycle is not really feasible and the RHVT does a very good job since it is robust and allows very efficient spot cooling.

Other applications might wait to be discovered. For instance, if spot cooling is needed in special corrosive environments, compressed nitrogen, carbon dioxide or argon as working gases in a RHVT can provide a good solution. Using medically clean working gases might be a solution for medical applications that require a sterile environment.

A, Kaufmann, *The Ranque Hilsch Vortex Tube Demystified*,
https://doi.org/10.1007/978-3-030-89766-6_6

Bibliography

1. H. Altenbach, *Kontinuumsmechanik* (Springer, Berlin, 2015)
2. R. Aris. *Vectors, Tensors and the Basic Equations of Fluid Mechanics* (Dover, Illinois, 1962)
3. R.T. Balmer, Pressure-driven Ranque-Hilsch temperature separation in liquids. J. Fluids Eng. **110**(2), 161–164 (1988)
4. J.M. Burazer, A.S. Cocic, M.R. Lecic, Numerical research of the compressible flow in a vortex tube using OpenFoam software. Thermal Sci. **20**, 1–14 (2016)
5. S. Candel. *Méecanique des Fluides* (Dunod, Paris, 1995)
6. P. Chassaing, *Mécanique des Fluides* (Céepaduès-Éditions, 1997)
7. P. Chassaing, *Turbulence en mécanique des Fluides* (Cépaduès-Éditions, 2000)
8. K.D. Devade, A. Pise, Parametric review of Ranque-Hilsch vortex tube. Am. J. Heat Mass Transf. **4**, 115–145 (2017)
9. S. Eiamsa-ard, P. Promvonge, Review of Ranque-Hilsch effects in vortex tubes. Renew. Sustain. Energy Rev. **12**(7), 1822–1842 (2008)
10. T. Farouk, B. Farouk, Large eddy simulation of the flow field and temperature separation in the Ranque-Hilsch vortex tube. Int. J. Heat Mass Transf. **50**, 4724–4735 (2007)
11. T. Farouk, B. Farouk, A. Gusol, Simulation of gas species and temperature separation in the counter-flow Ranque-Hilsch vortex tube using the large eddy simulation technique. Int. J. Heat Mass Transf. **52**, 3320–3333 (2009)
12. R. Hilsch, Die Expansion von Gasen im Zentrifugalfeld als Kälteprozeß, in *Zeitschrift für Naturforschung*. 1st ser. (1946), pp. 208–214
13. M. Kurosaka, Acoustic streaming in swirling flow and the Ranque-Hilsch (vortex-tube) effect. J. Fluid Mech. **124**, 139–172 (1982)
14. J. Mischner, V.I. Bespalov, Zur Entropieproduktion im Ranque-Hilsch Rohr. Forschung im Ingenieurwesen **67**, 1–10 (2002)
15. OpenCFD Ltd 2019. *Open Foam*. 2019. https://openfoam.com/
16. Sh.A. Piralishvili, A.A. Fuzeeva, Similarity of the energy seperation process in vortex Ranque tubes. J. Eng. Phys. Thermodyn. **79**, 27–32 (2006)
17. J.G. Polihronov, A.G. Straatman, Thermodynamics and angular propulsion in fluids. Phys. Rev. Lett. **109**, 054504 (2012)
18. H. Schlichting, *Boundary-Layer Theory* (Springer, Berlin, 1979)
19. R. Shamsoddini, A.H. Nezhad, Numerical analysis of the effects of nozzle number on the flow and power of cooling of a vortex tube. Int. J. Refrigeration **33**, 774–782 (2010)
20. T.K. Sharma, G.A.P. Rao, K.M. Murthy, Numerical analysis of a vortex tube: a review. Arch. Comput. Methods Eng. **24**, 251–280 (2017)

A, Kaufmann, *The Ranque Hilsch Vortex Tube Demystified*,
https://doi.org/10.1007/978-3-030-89766-6

21. K. Stephan et al., An investigation of energy separation in a vortex tube. Int. J. Heat Mass Transf. **26**(3), 341–348 (1983)
22. S. Subudhi, M. Sen, Review of Ranque-Hilsch vortex tube experiments using air. Renew. Sustaina. Energy Rev. **52**, 172–178 (2015)
23. J.J. Van Deemter, On the theory of the Ranque-Hilsch cooling effect. Appl. Sci. Res. Sect. A **3**, 174–196 (1952)
24. A.B. Vyas, J. Majdalani, Exact solution of the Bidirectional vortex. AIAA J. **44**(10), 2208–2216 (2006)
25. Y. Xue, M. Arjomandi, R. Kelso, A critical review of temperature seperation in a vortex tube. Exp. Thermal Fluid Sci. **34**, 1367–1374 (2010)
26. J. Zierep, *Similarity Laws and Modeling* (Marcel Dekke, New York, 1971)
27. J. Zierep, *Theoretische Gasdynamik* (Braun, Kronberg im Taunus, 1991)

The manufacturer's authorised representative in the EU is Springer Nature Customer Service Centre GmbH, Europaplatz 3, 69115 Heidelberg, Germany. If you have any concerns regarding our products, please contact ProductSafety@springernature.com

Printed and bound by CPI Group (UK) Ltd, Croydon, CR0 4YY
30/09/2026
02225710-0001